Practical Astronomy

Springer

London
Berlin
Heidelberg
New York
Barcelona
Budapest
Hong Kong
Milan
Paris
Santa Clara
Singapore
Tokyo

Other titles in this series

Photo-guide to the Constellations

A Self-Teaching Guide to Finding Your Way Around the Heavens

Chris Kitchin

Springer

Professor Chris Kitchin, BA, BSc, PhD, FRAS
University of Hertfordshire, University Observatory,
Bayfordbury, Hertfordshire, UK

ISBN 3–540–76203–5 Springer-Verlag Berlin Heidelberg New York

British Library Cataloguing in Publication Data
Kitchin, Christopher R. (Christopher Robert), 1947–
    Photo-guide to the constellations : a self-teaching guide
    to finding your way around the heavens. – (Practical
    astronomy)
    1.Constellations – Pictorial works
    I.Title
    523.8′0222
ISBN 3540762035

Library of Congress Cataloging-in-Publication Data
Kitchin, C. R. (Christopher R.)
    Photo-guide to the constellations : a self-teaching guide to
finding your way around the heavens / Chris Kitchin.
          p.   cm.
    Includes bibliographical references and index.
    ISBN 3–540–76203–5 (pbk. : alk. paper)
    1. Constellations – Observers' manuals.   2. Constellations –
Pictorial works – Handbooks, manuals, etc.   I. Title.
QB63.K47   1997   97–29324
523.8′022′2–dc21   CIP

Typeset by EXPO Holdings, Malaysia
Printed and bound by Interprint Limited, Malta
58/3830–543210   Printed on acid-free paper

Preface

What is the fascination that constellations hold for people? There are probably as many different answers to that question as there are people. For many, though, the constellations are the stepping-off point into the fabulous, mind-bending discoveries and concepts of modern astronomy. For others it is their long and intriguing history that beckons. For some people the constellations provide the means for navigation and orientation over the surface of the Earth, and of course there are the millions who place some faith in horoscopes. But for most people the patterns in the sky are a beautiful part of their environment to be treasured alongside the forests, fields and rivers that make life worth living.

However just as we are losing our green environment to pollution, so we are losing our sky. The glow from cities across the world swamps the stars in the night sky. Astronomers have had to retreat to remote mountain tops to escape that light pollution. The rest of us must make do with what is available. From the centre of a city, or any other brightly lit area, probably no stars at all will be visible even on the clearest of nights. From the suburbs, the brighter stars should normally be seen. Further out, but still in a populated area, most of the constellations will be visible. But to see constellations now as they were when first studied and named in antiquity, you also will have to journey to the secluded parts of the Earth.

Many star maps, atlases, photographs etc. show all the stars potentially visible to the naked eye. Indeed, in many cases they may extend down to much fainter stars that require telescopes in order to be seen at all. This is quite confusing if you are trying to find your way around the sky for the first time. Most of the stars included on such views will simply not be visible. In this book, therefore, three photographs are shown for every part of the sky. The first shows the stars that you may expect to see from a typical urban suburb suffering from light pollution, and in many cases that is, in truth, very few (no photographs are included to show what could be seen from the centre of the city – that would just be an orange-yellow splurge with no stars at all). The second shows the stars visible from a reasonable site, such as you might reach by driving a few miles out of town and getting away from any nearby lights. The final one shows the splendour that should be visible.

Whatever your reason for an interest in the constellations, the purpose of this book is to provide you with the means whereby you can find your way around the sky, whether you just wish to be able to find the Southern Cross or the Pole Star, or whether you wish to learn every single constellation visible in your part of the heavens.

In addition, the positions of some of the brighter individual objects such as galaxies, remnants of exploding stars, cradles wherein new stars are forming etc., and which can be seen with binoculars or a small telescope, are listed. Where needed, a small amount of the background astronomy on these objects is included, but that is not the main purpose of this book, and there are plenty of other sources of information on these topics for the interested reader to pursue. A little background on the history of the constellation, the derivation of its names and other items of interest is also covered.

I wish you clear skies, success in your endeavours, and hope that you find the same wonderment and joy in looking at the sky that I do.

Chris Kitchin
Hertford, 1997

Acknowledgements

I would like to thank Robert Forrest for many valuable discussions. Our joint book, *Seeing Stars* (Springer-Verlag, 1997), may well be found useful by readers wishing to progress beyond the introduction to the night sky to be found here. I would also like to thank Jeremy Bailey for obtaining the photographs of the southern constellations, and Paul Martin for printing innumerable apparently blank negatives with unfailing good humour.

Contents

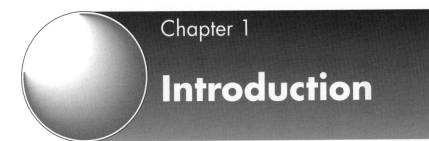

Chapter 1

Introduction

1.1 Starting Off

The purpose of this book is to provide you with the means whereby you may learn to find your way around the sky. That cannot however be accomplished just by reading this book. You will also need to go outside and look up at the night sky on many occasions with this book as a companion. Over the whole sky there are 88 constellations, but except near the equator, some of these will never be visible; they will always be below the horizon. Many of the constellations are formed from the fainter stars, and are not easily seen. Learning the sky does not therefore require you to recognise all 88 constellations. From most sites, the 15 or 20 constellations formed from the brighter stars will be quite sufficient. Once you know those, the remaining fainter constellations can soon be found.

The main guide to the constellations will be found in Chapter 2, and if you already know one or two constellations you may well wish to skip directly to that chapter. However, there are some "tricks of the trade", and most people will therefore find it useful to read this chapter before trying to recognise anything in the sky.

1.2 The Sizes of the Constellations

Probably the single most important impediment to recognising the constellations is a failure to realise just how BIG they are. This is a fault of the way that they are depicted in this and every other book, star chart or star atlas. But it is an inevitable fault. If the photographs in this book were to show the correct sizes of the constellations as they appear in the sky, then they would have to be about 20×30 inches (0.5×0.75 m) across when held at arm's length. Each photograph covers about a quarter of the visible hemisphere.

Looking at the images can therefore lead you to expect the constellations to be quite small when you search for them in the sky. They are not. The major constellations cover tens of degrees and stretch across large parts of the sky. To get an idea of the sizes that you should expect, first of all use the scale provided with the photographs to estimate the height and width of the constellation. Then use the useful guide that for most people the clenched fist held at arm's length (see Fig. 1.1, *overleaf*) covers about 8°, to step out the amount of sky that you should expect the constellation to cover. You will then at least start with the correct expectation of the size of the thing you are trying to find.

If there is a planetarium in your locality, then a visit will give an excellent start to learning the constellations, and more importantly will give you an idea of the scale of the patterns being sought.

1.3 The Lines

On the diagrams accompanying the sky photographs in Chapter 2, many of the stars are joined by lines. The reason for this is that vision does not occur through the

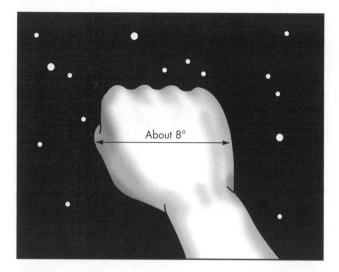

Figure 1.1. The clenched fist at arm's length spans about 8°.

eyes acting as optical instruments, but is the result of the eyes and brain acting in combination. We thus find it much easier to recognise patterns than individual dots.

The lines are therefore used to indicate patterns that you may be able to pick out in the sky. They have no significance other than that, and of course do not appear in the sky. The lines link the brighter stars in each constellation. There are clearly many different ways in which such patterns may be drawn, and the ones used here may well differ from those used in other sources. You may find that those other patterns suit you better than the ones shown here, or you may prefer to develop your own patterns. In all cases, the pattern to look for in the sky is the one that suits you best.

1.4 Star Hopping

Once you have found and learnt one constellation, then the surrounding ones can usually be found quite easily. This process is called *star hopping*, and it is the basis of how you find your way around the sky. Two bright stars in the known constellation are selected so that the line joining them, when extended (sometimes a slight curve is needed) meets a bright star in the next constellation. Having reached one star in the next constellation, the pattern of the rest of that constellation can usually be

picked out without difficulty. Some such hops are very well known. The two end stars of the constellation Ursa Major, for example, are called the pointers because they are used so frequently to find the pole star (Polaris – Figs 1.2, 2.13 and 2.22).

1.5 Dark Adaption

In order to see as many stars as possible, it is essential to let your eyes adapt to the dark. This is a physiological process whereby the light-sensitive molecules in the retina increase dramatically in number under low-light level conditions. The eye's sensitivity also increases correspondingly. Additionally the pupil of the eye becomes larger in the dark, allowing more light into the eye, but it is the increase in the number of light-sensitive molecules that causes the greatest improvement in sensitivity. The light-sensitive molecules take about 20 to 30 minutes to regenerate after being in a brightly lit area, and your ability to see in the dark accordingly improves over the same period. Most people will be familiar with the

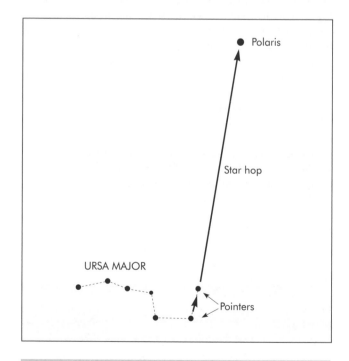

Figure 1.2. Star hopping from the pointers in Ursa Major to the Pole Star (Polaris).

phenomenon, and it is also known as night vision. Upon walking outside at night from a brightly illuminated room, at first very little can be discerned. However quite quickly you will begin to see things around you, and after half an hour or so, you can probably see quite well.

Dark adaption therefore simply consists of being in and STAYING in the dark for 20 minutes or more. Bright light, even using a torch or lighting a cigarette, will remove some or all of your dark adaption in seconds, and you then require another half hour wait for its return.

There is thus a problem: you will probably need to look at the diagrams and photographs in this book while you a trying to identify the constellations. But light bright enough to enable you to read the book will almost certainly reduce your dark adaption. There are two ways around the problem. The first is that the brighter stars in the major constellations should be visible even if your dark adaption is poor. So to start with, while learning the major constellations, you can illuminate the book well enough to read it, and still be able to spot the bright stars. Later when you know the main outlines, it will not be too difficult to memorise the fainter stars and constellations, and so find them with dark-adapted eyes. The second approach relies on the property of the eye that dark adaption is not affected by deep red light. Thus a red filter over your torch will enable you to work without losing dark adaption. However, to be effective, the filter must be a very deep red. Some of the deep red filters sold as safe lights for use in photographic dark rooms are suitable. Cheaper gelatine or plastic type filters are normally not red enough, although they are better than white light.

1.6 The Moon, Planets and other Problems

For anyone observing the sky with a telescope, the Moon and planets are among the most favoured sights. However they are something of a nuisance when you are learning the constellations.

The Moon, especially from half through full and back to half again, is so bright that, even on the clearest of nights, the fainter stars will be swamped by its scattered light. On hazy nights, there may not be a star to be seen with the naked eye. Thus in addition to the normal problems of cloudy nights, light pollution etc., the presence of a bright Moon in the sky means that for another 50 per cent of the time, the fainter stars and constellations will not be seen.

There is little to be done about this except to reserve your attempts to find those fainter stars and constellations for the moonless nights. One "advantage" of a bright Moon is that your eyes will not reach full dark adaption, and so the use of a torch or other light to allow you to read the diagrams and photographs in this book while trying to spot the patterns in the sky will not matter so much.

The planets do not have the devastating effect upon visibility caused by a bright Moon, but they can lead to confusion. To the naked eye they appear like bright stars, and their positions change noticeably over periods from a few days to a few weeks. The presence of a planet within a constellation will change the pattern of the constellation and make it much more difficult to recognise.

However in practice the confusion caused by planets is not as great as it might be. Firstly, the paths of the planets around the sky are confined to the zodiac (Section 1.8). The 76 constellations not within the zodiac will therefore never have a planet within their confines. Secondly, of the eight planets, only two are really likely to be a problem. Uranus, Neptune and Pluto are not visible to the naked eye. (In theory, Uranus is just visible to the naked eye; it is however far fainter than any of the stars used here to depict the constellations, and is not likely to be a problem.) Mercury, although quite bright, is always close to the Sun, and if visible at all, it is only just after sunset, or just before sunrise. Few of the stars or constellations are therefore likely to be visible in its vicinity. Venus and Jupiter are always brighter than any of the stars. Venus ranges from 8 to 17 times brighter than Sirius (in Canis Major – the brightest star in the night sky), and Jupiter from 1.5 to 3 times brighter. Thus only Mars and Saturn are really likely to be a problem, and Mars has a noticeable red colour that distinguishes it from almost all the stars other than Antares (in Scorpius).

The positions of the planets can be found from charts published in the popular astronomy magazines, and in many of the quality newspapers. If you think that there is an interloper in the (zodiacal) constellation that you are trying to find, then a quick check in such sources will tell you if it is likely to be a planet.

Other objects that can change the patterns of the constellations include novae, supernovae, spacecraft and aeroplanes. Novae and supernovae are exploding stars; many occur every year, but those bright enough to be seen by the naked eye only about once a decade. So these are not too much of a problem. Spacecraft appear star-like to the naked eye, and there are many thousands in orbit around the Earth. Perhaps one or two hundred spacecraft are bright enough to cause potential confusion while recognising constellations. However all such brighter spacecraft are in low orbits, and will be seen to move in a matter of seconds. They will typically take 10 to 15 minutes to cross the whole sky. Aircraft will also normally be moving. From some sites though, you may be looking directly along a fight path, and aircraft on that flight path will then appear stationary in the sky, sometimes for periods of several minutes. There are also likely to be a series of aircraft so that as one disappears another takes its place. If you continue watching such "stars" they will eventually move, revealing their true identity. A pair of binoculars may also be a help, since they may enable you to see the other lights of the aircraft. After a short while, you will learn where the flight path is in the sky, and the aircraft on it will no longer cause a problem.

1.7 The Names of the Stars

There are many ways of naming the stars, and some stars have several names. However, one way that is not possible, except for a few exceptions, such as Barnard's star, where an unusual star has been named after its discoverer, is after people. A frequently encountered and cruel swindle is based upon this. Advertisements may be found from time to time in quite reputable sources from firms offering to name stars after individuals for a charge. This is not illegal because they will dutifully write your name (or your friend's or relative's) after receiving your money against the selected star in a star catalogue or on a star map. That name however will not be officially recognised and no one will ever use it. Most stars are correctly identified simply by their numbers in a catalogue. The brighter stars, which include the ones used here to define the constellations, are named on the Bayer system. In addition some have individual names such as Sirius, Polaris and Antares.

Under the Bayer system which was first used in the *Uranometria* star catalogue of 1603, the brighter stars within a constellation are given Greek-letter labels followed by the abbreviation of the constellation name. Usually, this is in order of their brightnesses, so that α designates the brightest star of the constellation, β, the second brightest and so on (the full Greek alphabet is listed in Appendix 2 and the abbreviations used for the constellations in Section 1.10). The system however breaks down at times as will be seen in Chapter 2, with for

example, δ UMa being fainter than ε UMa, ζ UMa and η UMa. In other cases, some of the Greek letters may be missing. Thus there was once a large southern constellation called Argo, but during the eighteenth century this was split into Carina, Puppis, Pyxis and Vela, but without reassigning Greek letters. Thus the brightest stars in Vela, for example, are labelled γ Vel, δ Vel and λ Vel.

There are 24 letters in the Greek alphabet, and once these have been utilised, then the Bayer system continues with the letters, a, b, c … . When those letters have been used up, the letters, A, B, C, … up to Q are used. The letters R to Z are reserved for variable stars.

After an individual name, or a Bayer designation, the star's Flamsteed number, from his *Historia Coelestis* catalogue of 1725, is usually used, again followed by the abbreviation of the constellation name. Thus Pleione in the Pleiades is also known as 28 Tau. Most modern catalogues however ignore the constellations, and simply list stars in order of their positions in the sky. There are many such catalogues, some listing millions of stars. The star is then known from an abbreviation of the catalogue name, and its number in that catalogue. Thus Dubhe (α UMa, 50 UMa) is also called HD 95689 (from the Henry Draper catalogue), BS4301 (from the Bright Star catalogue), and BD +62° 1161 (from the Bonner Durchmusterung catalogue). For the purposes of recognising the constellations, however, the star's name or Bayer designation will be sufficient.

1.8 The Zodiac

This is a band of the sky about 18° wide and centred upon the Sun's annual path around the sky. (The Sun's annual path around the sky is called the *ecliptic*. The Sun's movement is of course actually due to the Earth's motion around its orbit. The ecliptic is also therefore the plane of the Earth's orbit extended off into space.) The Moon and all the major planets except Pluto are always to be found within this band. The zodiac was divided by the ancient Greeks into twelve segments, each 30° long, and called the *signs of the zodiac*. The stars within each segment were then organised into a constellation which was given the same name as the sign of the zodiac. Over the last two and a half thousand years though, the stars have moved more than 30° eastwards in relation to the signs of the zodiac. The constellations therefore no longer match the signs of the zodiac. Additionally, the movement of the stars has caused the Sun's annual path around the sky to pass through a new constellation, Ophiuchus, though rather illogically this is not normally considered as a zodiacal constellation. The full list of constellations through which the Sun now passes is thus:

Constellation	Dates of solar passage
Sagittarius	Dec 19 to Jan 21
Capricornus	Jan 22 to Feb 16
Aquarius	Feb 17 to Mar 12
Pisces	Mar 13 to Apr 18
Aries	Apr 19 to May 14
Taurus	May 15 to June 21
Gemini	June 22 to July 21
Cancer	July 22 to Aug 11
Leo	Aug 12 to Sept 17
Virgo	Sept 18 to Oct 31
Libra	Nov 1 to Nov 22
Scorpius	Nov 23 to Nov 30
Ophiuchus	Dec 1 to Dec 18

(For comparison, the Sun is in the zodiacal *sign* of Sagittarius from 22 Nov to 21 Dec.)

1.9 The Milky Way, Magellanic Clouds and Zodiacal Light

The Milky Way is what we can see of the huge collection of stars of which our Sun is a member and which is just one of the many millions of galaxies throughout the universe. There are perhaps 100,000,000,000 stars in the Milky Way galaxy, but all except a few thousand are too faint to see individually with the naked eye. Nonetheless the accumulated total from millions of very faint stars can be seen. Since the galaxy has a wheel-like shape, and we are embedded within it towards one edge, we see the other stars as a faint band of light circling the whole sky. The Milky Way is easily visible on a clear dark night from a good site, but will disappear into the background from a light-polluted site, or when there is a bright Moon around. It runs mainly through the following constellations, though it may also overlap slightly into the ones on either side:

Cassiopeia
Perseus

Auriga
Orion/Gemini
Monoceros

Puppis
Vela
Centaurus/Crux/Musca
Circinus
Norma

Ara/Scorpius
Sagittarius
Scutum
Aquila
Sagitta

Vulpecula
Cygnus
Lacerta/Cepheus

The Magellanic clouds, also known as Nubecula Major (the LMC) and Nubecular Minor (the SMC), are named for Ferdinand Magellan who led the first circumnavigation of the Earth (1519 to 1521), though killed during the voyage), and are two much smaller galaxies which are satellites of our own Milky Way galaxy; they are visible in the southern hemisphere as faint glowing patches of light. The large Magellanic cloud is found on the border of Dorado and Mensa, the small Magellanic cloud within Tucana.

The zodiacal light is a faint glow that follows the line of the zodiac. It is most prominent near the horizon following sunset, or before sunrise, but can be traced all around the sky, and is due to light from the Sun being scattered back towards us by dust particles between the planets. Like the Milky Way and the Magellanic clouds, it can be seen from a good site on a moonless night without difficulty but soon fades if there is any background glow.

1.10 The Constellations

The full list of the modern constellations, their abbreviations, genitives, and where they are to be found in the photographs and diagrams in Chapter 2 is given in Table 1.1.

Table 1.1. The modern constellations

Constellation	Abbreviation	Genitive	English Meaning*	Location in Chapter 2**
Andromeda	And	Andromedae	Female Name	2.22, <u>2.24</u>, <u>2.30</u>, 2.37, 2.62, 2.111
Antlia	Ant	Antliae	Pump	<u>2.117</u>, 2.122
Apus	Aps	Apodis	Bee	<u>2.76</u>, <u>2.82</u>, <u>2.88</u>
Aquarius	Aqr	Aquarii	Water Carrier	<u>2.30</u>, 2.88, 2.94, 2.111
Aquila	Aql	Aquilae	Eagle	2.37, <u>2.43</u>, <u>2.94</u>
Ara	Ara	Arae	Altar	<u>2.82</u>
Aries	Ari	Arietis	Ram	<u>2.24</u>, 2.30, 2.62, 2.111
Auriga	Aur	Aurigae	Charioteer	2.22, <u>2.24</u>, <u>2.62</u>, 2.69
Boötes	Boo	Boötis	Herdsman	<u>2.13</u>, 2.43, 2.48
Caelum	Cae	Caeli	Chisel	<u>2.100</u>
Camelopardalis	Cam	Camelopardalis	Giraffe	<u>2.22</u>, <u>2.24</u>, <u>2.62</u>
Cancer	Cnc	Cancri	Crab	<u>2.55</u>, <u>2.69</u>, 2.122
Canes Venatici	CVn	Canum Venaticorum	Hunting Dogs	<u>2.13</u>, 2.48
Canis Major	CMa	Canis Majoris	Large Dog	2.69, <u>2.100</u>, <u>2.117</u>
Canis Minor	CMi	Canis Minoris	Small Dog	<u>2.55</u>, <u>2.69</u>, 2.117
Capricornus	Cap	Capricorni	Goat	2.43, 2.88, <u>2.94</u>, 2.111
Carina	Car	Carinae	Keel	2.76, <u>2.100</u>, <u>2.117</u>
Cassiopeia	Cas	Cassiopeiae	Female Name	<u>2.22</u>, <u>2.24</u>, 2.62
Centaurus	Cen	Centauri	Centaur	<u>2.76</u>, 2.82, 2.100, 2.117
Cepheus	Cep	Cephei	Male Name	<u>2.22</u>, 2.37
Cetus	Cet	Ceti	Whale	<u>2.106</u>, <u>2.111</u>
Chamaeleon	Cha	Chamaelontis	Chamaeleon	2.76, <u>2.88</u>, <u>2.100</u>
Circinus	Cir	Circini	Compass	<u>2.76</u>, <u>2.82</u>
Columba	Col	Columbae	Dove	<u>2.100</u>, 2.117
Coma Berenices	Com	Comae Berenices	Berenice's Hair	<u>2.13</u>, 2.48, <u>2.55</u>
Corona Australis	CrA	Coronae Australis	Southern Crown	<u>2.82</u>, <u>2.94</u>
Corona Borealis	CrB	Coronae Borealis	Northern Crown	<u>2.13</u>, <u>2.43</u>, <u>2.48</u>
Corvus	Crv	Corvi	Crow	2.76, <u>2.117</u>, <u>2.122</u>
Crater	Crt	Crateris	Cup	<u>2.117</u>, <u>2.122</u>
Crux	Cru	Crucis	Cross	<u>2.76</u>, <u>2.117</u>
Cygnus	Cyg	Cygni	Swan	2.22, 2.30, <u>2.37</u>, 2.43
Delphinus	Del	Delphini	Dolphin	<u>2.30</u>, <u>2.37</u>, 2.94
Dorado	Dor	Doradus	Swordfish	2.88, <u>2.100</u>
Draco	Dra	Draconis	Dragon	(<u>2.13 & 2.37</u>), 2.22, 2.62
Equuleus	Equ	Equulei	Little Horse	<u>2.30</u>, <u>2.37</u>, <u>2.94</u>
Eridanus	Eri	Eridani	River	(<u>2.88 & 2.106</u>), 2.69, 2.100
Fornax	For	Fornacis	Furnace	<u>2.106</u>
Gemini	Gem	Geminorum	Twins	2.55, 2.62, <u>2.69</u>
Grus	Gru	Gruis	Crane	<u>2.82</u>, <u>2.88</u>, 2.94, 2.111
Hercules	Her	Herculis	Male Name	2.37, 2.43, <u>2.48</u>, 2.94
Horologium	Hor	Horologii	Clock	(<u>2.88 & 2.100</u>)
Hydra	Hya	Hydrae	Sea Serpent	2.55, 2.69, 2.76, 2.117, <u>2.122</u>
Hydrus	Hyi	Hydri	Little Snake	2.82, <u>2.88</u>, 2.100
Indus	Ind	Indi	Indian	<u>2.82</u>, <u>2.88</u>
Lacerta	Lac	Lacertae	Lizard	<u>2.22</u>, 2.30, <u>2.37</u>
Leo	Leo	Leonis	Lion	2.13, <u>2.55</u>, <u>2.122</u>
Leo Minor	LMi	Leonis Minoris	Little Lion	<u>2.13</u>, <u>2.55</u>, <u>2.122</u>

Table 1.1. (Continued)

Constellation	Abbreviation	Genitive	English Meaning*	Location in Chapter 2**
Lepus	Lep	Leporis	Hare	2.69, 2.100, 2.106
Libra	Lib	Librae	Scales	2.43, 2.48, 2.76
Lupus	Lup	Lupi	Wolf	2.76, 2.82
Lynx	Lyn	Lyncis	Lynx	2.55, 2.62, 2.122
Lyra	Lyr	Lyrae	Lyre	2.37, 2.43
Mensa	Men	Mensae	Table	2.88, 2.100
Microscopium	Mic	Microscopii	Microscope	2.82, 2.88, 2.94
Monoceros	Mon	Monocerotis	Unicorn	2.55, 2.69, 2.100, 2.117
Musca	Mus	Muscae	Fly	2.76, 2.100
Norma	Nor	Normae	Rule	2.76, 2.82
Octans	Oct	Octantis	Octant	2.76, 2.82, 2.88
Ophiuchus	Oph	Ophiuchi	Serpent Carrier	2.43, 2.48, 2.82, 2.94
Orion	Ori	Orionis	Male Name	2.69, 2.100, 2.106
Pavo	Pav	Pavonis	Peacock	2.82, 2.88
Pegasus	Peg	Pegasi	Flying Horse	2.24, 2.30, 2.37, 2.94, 2.111
Perseus	Per	Persei	Male Name	2.22, 2.24, 2.62
Phoenix	Phe	Phoenicis	Phoenix	2.88, 2.111
Pictor	Pic	Pictoris	Painter	2.100
Pisces	Psc	Piscium	Fishes	2.24, 2.30, 2.106, 2.111,
Piscis Austrinus	PsA	Piscis Austrini	Southern Fish	2.88, 2.94, 2.111
Puppis	Pup	Puppis	Stern	2.100, 2.117
Pyxis	Pyx	Pyxidis	Mariner's Compass	2.100, 2.117
Reticulum	Ret	Reticuli	Net	2.88, 2.100
Sagitta	Sge	Sagittae	Arrow	2.37, 2.43, 2.94
Sagittarius	Sgr	Sagittarii	Archer	2.43, 2.82, 2.94
Scorpius	Sco	Scorpii	Scorpion	2.43, 2.48, 2.76, 2.82, 2.94
Sculptor	Scl	Sculptoris	Sculptor	2.88, 2.106, 2.111
Scutum	Sct	Scuti	Shield	2.43, 2.94
Serpens	Ser	Serpentis	Serpent	2.43, 2.48, 2.94
Sextans	Sex	Sextantis	Sextant	2.55, 2.122
Taurus	Tau	Tauri	Bull	2.24, 2.62, 2.69, 2.106
Telescopium	Tel	Telescopii	Telescope	2.82
Triangulum	Tri	Trianguli	Triangle	2.24, 2.30, 2.62, 2.111
Triangulum Australe	TrA	Trianguli Australis	Southern Triangle	2.76, 2.82, 2.88
Tucana	Tuc	Tucanae	Toucan	2.82, 2.88
Ursa Major	UMa	Ursae Majoris	Great Bear	2.13, 2.22, 2.55, 2.62
Ursa Minor	UMi	Ursae Minoris	Little Bear	2.13, 2.22
Vela	Vel	Velorum	Sails	2.76, 2.100, 2.117
Virgo	Vir	Virginis	Virgin	2.13, 2.43, 2.48, 2.55, 2.76, 2.122
Volans	Vol	Volantis	Flying Fish	2.88, 2.100, 2.117
Vulpecula	Vul	Vulpeculae	Fox	2.37, 2.43

* For further details see Chapter 3.
** The figure numbers refer to the diagrams showing all the constellations contained in an image. Figures containing all or nearly all of the main stars of a constellation are underlined. In a few cases two images are required to encompass the whole constellation, and then those two images are bracketed and underlined. The photographs showing appearances of that area of sky under various observing conditions, and star hopping routes will be found immediately prior to the figures listed.

Chapter 2

Finding the Constellations

2.1 Introduction

The fundamental method of finding your way around the sky is star hopping (Section 1.4). This requires that you can recognise one constellation in order to find another. Finding your first constellation is therefore rather different from finding your second or third. The first constellation, or starter constellation, must be easily recognisable in the sky without prior knowledge. Suitable starter constellations vary depending upon the time of year and your position on the Earth, and so this chapter is subdivided into sections for northern hemisphere observers (latitudes 20° to 90°), equatorial observers (latitudes –30° to 30°), and southern hemisphere observers (latitudes –90° to –20°).

Directions for finding the constellations are given here assuming that you are starting from scratch; that is, assuming that you cannot recognise a single constellation. If you can recognise one or more constellations already, then whether or not that is the starter constellation suggested here, the one you can find is the one to start with. You may also know someone who can show you some of the constellations to get you started; boy scouts and girl guides are useful for this. Alternatively, you could go along to your local astronomical society where you will undoubtedly find yourself deluged with help. But although such prior knowledge or outside help should not be spurned, it is not essential to finding even your first constellation.

As a reminder, since this is probably the single greatest problem in starting to learn your way around the sky, remember that constellations are BIG (Section 1.2). Always use the scale on the photographs to step out the size in the sky that is to be expected for the constellation before you search for it.

2.2 Northern Hemisphere Observers
(Latitudes 20° to 90°)

Fifty or sixty constellations are probably visible to observers in these latitudes, although many of them are small and faint. For a start the brighter or major constellations, of which there are seventeen, are sufficient to find your way around the sky.

2.2.1 The Major Constellations

The major constellations for northern hemisphere observers are listed below. There is, of course, a considerable overlap with those for equatorially based observers, and even some overlap with the major constellations for southern observers.

The Major Constellations for Northern Hemisphere Observers

Andromeda	Aquila	Auriga	Boötes
Canis Major	Canis Minor	Cassiopeia	Cygnus
Gemini	Leo	Lyra	Orion
Pegasus	Perseus	Taurus	Ursa Major
Ursa Minor			

The best constellation to start with is the Plough, also known as the Great Bear, Ursa Major, King Charles' Wain, the Big Dipper or, less officially, the Saucepan (strictly, the Plough is formed from just the seven main stars of Ursa Major). This is because for latitudes to the north of about 40°, the main stars of the constellation will always be visible on a clear night (that is, the stars are circumpolar – they never set). Even at the equator, some or all of the main stars of the Plough will be in the sky for 60 per cent of the time. It is often a puzzle why constellations have the names that they do, because the shapes outlined by the main stars rarely resemble the object they are supposed to represent. In many cases the name has no association with shape, but this is not the case with the Plough, though we have to imagine the ancient horse-drawn implement (Fig. 2.1) rather than the modern multi-tiller dragged by a tractor.

Ursa Major

How then do you find Ursa Major for the first time? One of the problems with finding any constellation you have

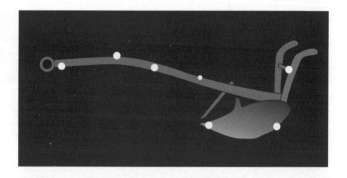

Figure 2.1. The Plough (Ursa Major).

never recognised before is that its appearance will change depending upon circumstances. Thus, from an urban observing site, probably only the seven main stars of the constellation will be visible (Figs 2.2 and 2.3 – *all photographs and their accompanying star maps in this chapter are shown to exectly the same scale and can therefore easily be compared*). In poor, hazy conditions and with a bright Moon, you may not be able to see any stars at all.

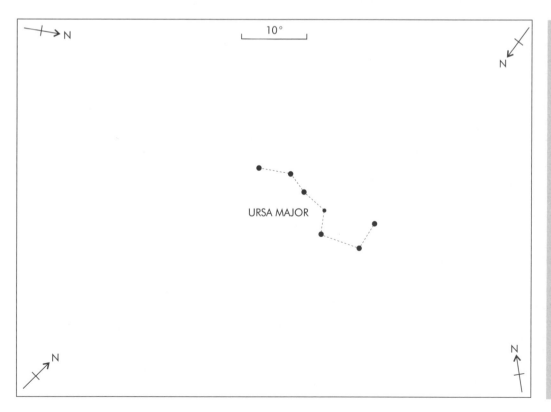

N   10°   N

URSA MAJOR

Magnitude scale on the star maps

● Magnitude 1.5+
● Magnitude 2.5–1.5
• Magnitude 3.5–2.5
· Magnitude 4.5–3.5
· Magnitude 4.5 and less

Figure 2.2. The seven main stars of Ursa Major – also called the Plough. (Note that the lines joining the stars on this and subsequent figures are only there to help fix the pattern of the constellation in your mind. They have no other significance.)

10°

N

Figure 2.3. The Ursa Major region – stars visible from a typical urban light-polluted site.

From a reasonable site, on a good clear night without a Moon, on the other hand, you will probably be able to see two or three dozen stars in the constellation (Fig. 2.4, *overleaf*). The appearance will also depend upon the orientation of the constellation (Fig. 2.5, *overleaf*). This is less variable for constellations near the equator, but for those near the pole, like Ursa Major, they can be at any angle. Finally, the constellation may be at different altitudes above the horizon, or from some latitudes some or all the stars may be below the horizon. For Ursa Major, the constellation will have the appearance of Fig. 2.5c when it is highest in the sky, that of Fig. 2.5a when it is lowest. Figure 2.5d shows it at its western-most position, and Fig. 2.5b at its eastern-most position.

With these problems in mind therefore, it is best to select a reasonably clear moonless night, and if need be to move away from town and other bright lights, for your first attempt at constellation identification. As you are able to recognise more and more constellations, you will find that it is not essential to have good observing conditions in order to find your way around the sky, but it will help considerably to begin with. You will also need to know the approximate direction of the north on the ground. If you do not know this already for your area,

then it can easily be obtained from a good-quality map. Many will have a compass rose marked, but if not, maps in the northern hemisphere are almost invariably oriented with north at the top.

Having chosen a reasonable night, arrived at a reasonable observing site, and oriented yourself to face north, what then? First get an idea of the area of sky over which you will need to search. Ursa Major is about 35° from the North Pole. You will therefore need to search over a circle extending from 35° either side of north, and from near the horizon to near the zenith. This radius is about four times the width of the clenched fist at arm's length (Fig. 1.1). The constellation's seven main stars extend over a distance of about 20° (2.5 × the clenched fist at arm's length). Step these two distances out so that you can visualise the scale of the constellation and of the search area. Then with the possible appearances in mind (Fig. 2.5) look among the brighter stars in the search area until you can find the saucepan-like pattern of the Plough (Fig. 2.6, *overleaf*). You should be able to spot the constellation within a minute or two.

If after ten or fifteen minutes searching you have not found the constellation, then you are probably not facing north, so check your orientation. Alternatively, if your

10°

N N N N

Figure 2.4. The Ursa Major region – stars visible from a good site.

latitude is south of about 40°, then some of the main stars of Ursa Major may be below your horizon (see Chapter 4, also astronomy magazines and some national newspapers publish monthly sky charts showing the stars visible for the appropriate time of year, so you can check these if you suspect that Ursa Major is below the horizon). In the latter case you can either try again at a different time of night or later in the year, or start with a different constellation (see below). If Ursa Major should be visible, and you are oriented correctly to see it, and yet you still cannot find it, then you will probably have to find someone who can point it out to you directly.

Assuming that you have been successful in finding the main stars of Ursa Major, you can then start to become familiar with their individual names. The seven bright stars of Ursa Major are known as Dubhe, Merak, Phad, Megrez, Alioth, Mizar and Alkaid (Fig. 2.7, *overleaf*). Mizar has a fainter close companion called Alcor that can be seen next to it if you have average or better eyesight. The seven main stars also have designations under the Bayer system as α Ursa Majoris (usually abbreviated as α UMa – see Section 1.10 for the genitives and abbreviations of all the constellation names), β UMa, γ UMa, δ UMa, ε UMa, ζ UMa and η UMa (Fig. 2.8, *overleaf*).

Once you are familiar with the main stars of Ursa Major, you should start looking for the fainter ones (Figs 2.4, 2.9 and 2.12).

As well as the naked-eye stars, each constellation is now used to delimit a specific area of the sky. These areas cover the whole sky and are officially recognised by the International Astronomical Union (see, for example, the outer boundary of Ursa Major marked in Fig. 2.9, *overleaf*). Since the boundaries were designated in 1922, and most stars were named before that date, there are numerous anomalies in the system. Thus within the official boundaries of Ursa Major may be found the stars 1 CVn (Canes Venatici), 55 Cam (Camelopardalis) and 15 LMi (Leo Minor); while 10 UMa is actually in the area assigned to the constellation Lynx, etc. Fortunately, both the official boundaries and these latter complications are something you can completely ignore while learning your way around the sky.

Ursa Minor

Once you can identify Ursa Major (or any other "starter" constellation), finding other constellations becomes much

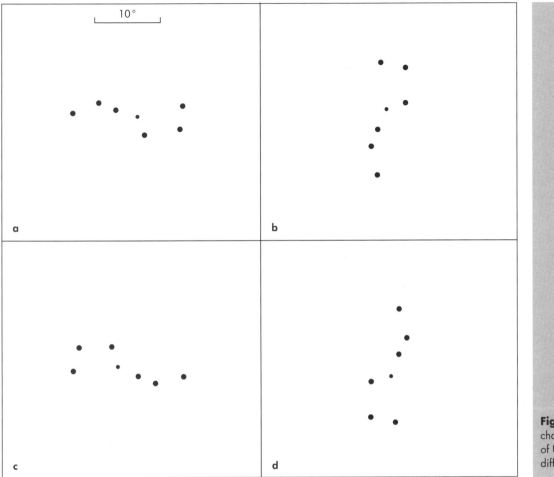

Figure 2.5. The changing appearance of Ursa Major in different orientations.

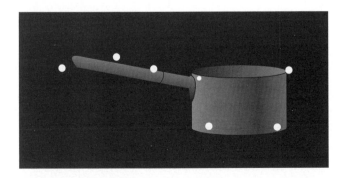

Figure 2.6. Ursa Major as a Dipper or Saucepan.

easier. Ursa Major is a good starter constellation because it provides helpful guides to its surrounding constellations. The end stars (α and β UMa) indeed are often called the *pointers*, because they may be used to find the Pole Star (α Ursa Minoris, Polaris – Fig. 1.2 and Fig. 2.10, *overleaf*). The Pole Star is a reasonably bright star that by chance happens to be very close to the position of the North Pole in the sky. It is therefore very useful as a guide to finding your orientation on the surface of the Earth. To find Polaris simply follow the line of the pointers for about five times their separation, and you come directly to the star (Fig. 2.10). Polaris is the brightest star in the constellation Ursa Minor, and so the rest of that constellation can be easily identified once Polaris has been recognised Figs 2.4, 2.10 and 2.13).

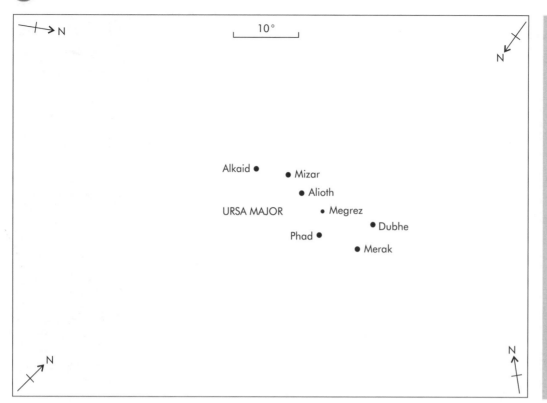

Figure 2.7. The names of the main stars in Ursa Major.

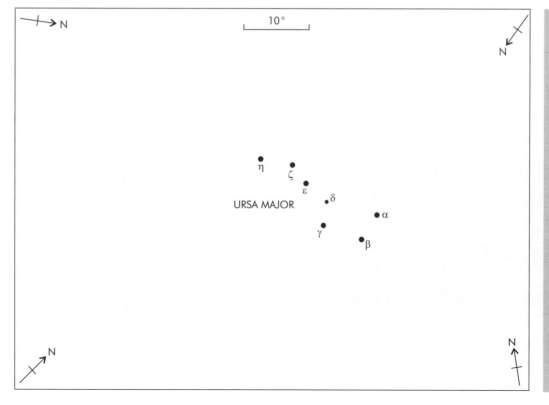

Figure 2.8. The Bayer designations of the main stars in Ursa Major.

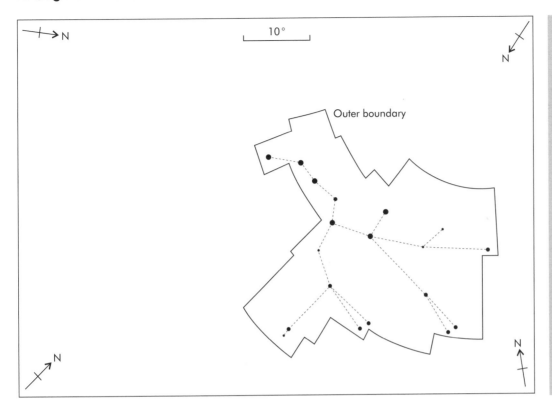

Figure 2.9. Ursa Major showing the stars usually visible to the naked eye from a good site, plus its official outer boundary.

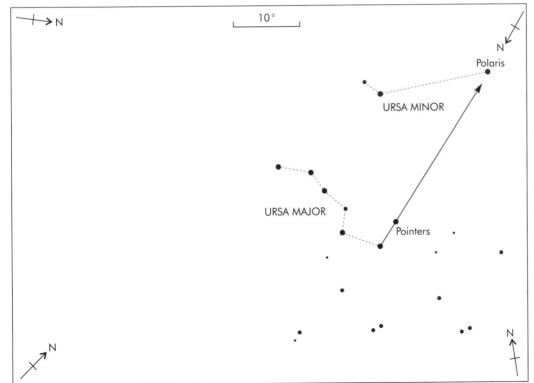

Figure 2.10. Star hopping from the pointers in Ursa Major (α UMa and β UMa, or Dubhe and Merak) to the Pole Star (α UMi or Polaris) and thence to Ursa Minor.

Boötes

The "tail" stars of Ursa Major give the direction to α Boö (Fig. 2.11). Follow them around in a curving line for about five times the separation of the last two stars (ζ UMa and η UMa) to the bright, slightly reddish star, α Boö (Arcturus). This is the fourth brightest star in the night sky and the rest of the constellation of Boötes, which however is much fainter than Arcturus, then follows (Fig. 2.11).

Cassiopeia

On the far side of the pole from Ursa Major is another good "starter" constellation, Cassiopeia. It is about the same distance from the pole as Ursa Major, a little smaller in extent, and easily recognisable. If you cannot find Ursa Major as a starting point, or it is below the horizon, then since Cassiopeia is on the opposite side of the pole it should be easily visible. This is a constellation whose shape bears no resemblance to its mythological image (the figure of a human female), but it is easily recognisable as an "M", "Σ" or "W" (depending on its orientation) in the sky. It may be found by star hopping from γ UMa to α UMi and then carrying on in a straight line for almost the same distance again (Figs. 2.14 and 2.15, *overleaf*). Since the constellation is close to the North Pole, its appearance will change with time (Fig. 2.16, *overleaf*). This star hopping brings you to β Cas, and the rest of the constellation may then be found (Figs 2.17 to 2.20 inclusive, *overleaf*).

Andromeda

β Cas and α Cas then give a straight line star hop over about five times their separation to γ And and so to the rest of the constellation of Andromeda (Fig. 2.21, *overleaf*). The great galaxy in Andromeda, M31 (NGC 224), may be seen with the naked eye on a good night about 7° north-west of β And (Fig. 2.21).

Andromeda may perhaps be more easily found as part of a curving line of five equally bright stars including α Per and β Peg, with equal separations between them (Figs 2.23 and 2.24, *overleaf*). Groupings of stars like these five, which are not recognised constellations, are sometimes called *asterisms*. This line of stars covers a large angle, but is quite easy to distinguish.

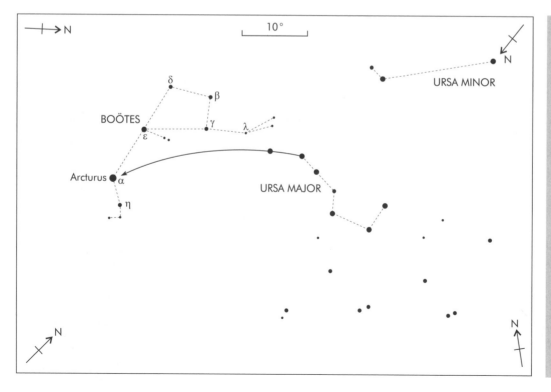

Figure 2.11. Star hopping from ζ UMa and η UMa (Mizar and Alkaid) to Arcturus (α Boö) along a slightly curving path and then to the rest of Boötes.

Figure 2.12. The Ursa Major region – the stars visible from a brilliant site to an acute observer. These most detailed photographs of each region of the sky show at least all the stars visible to the naked eye under the best conditions.

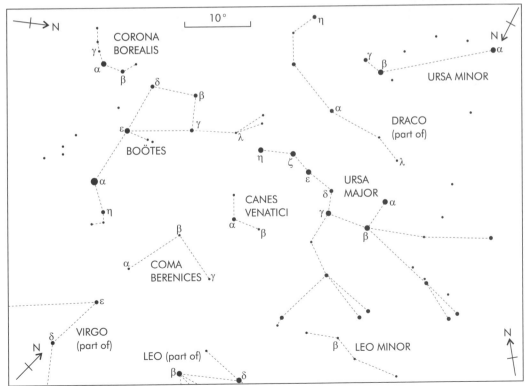

Figure 2.13. All the constellations in the Ursa Major region.

Figure 2.14. Finding Cassiopeia – stars visible from a typical urban, light-polluted site.

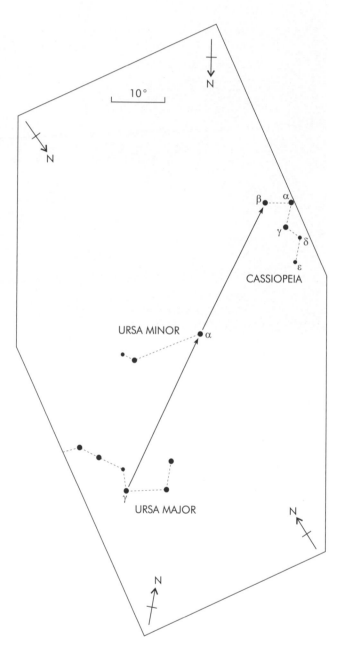

Figure 2.15. Star hopping from Ursa Major and Ursa Minor to Cassiopeia.

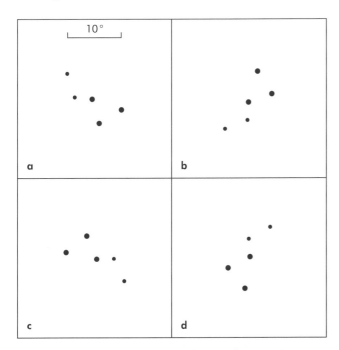

Figure 2.16. The changing appearance of Cassiopeia in different orientations: **a** lowest in the sky; **b** eastern-most in the sky; **c** highest in the sky; **d** western-most in the sky.

Pegasus

The line ends in the star β Peg, and this provides the next star hop to the constellation of Pegasus (Figs 2.25 and 2.26). The *Square of Pegasus* is easily recognisable in the sky and forms another useful starter constellation. It is however something of a misnomer, since the square requires α And for its completion. Pegasus is a large constellation shown in its entirety in Figs 2.27 and 2.28.

Cygnus

From Pegasus a straight line star hop from β through γ Peg takes us to γ Cyg, and the rest of Cygnus (Figs 2.31, 2.32 and 2.33).

Lyra

Then from Cygnus we may hop to α Lyr (Vega) in a slightly curving line from α Cyg (Deneb) through δ Cyg

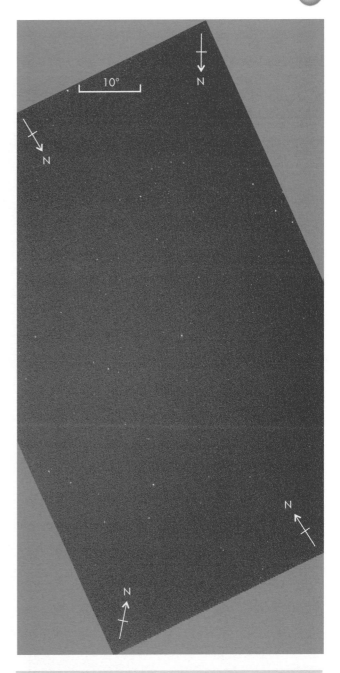

Figure 2.17. Finding Cassiopeia – stars visible from a good site.

(Fig. 2.34). The remaining stars in Lyra are much fainter than Vega, but may be seen on a good night without difficulty (Figs 2.36 and 2.37).

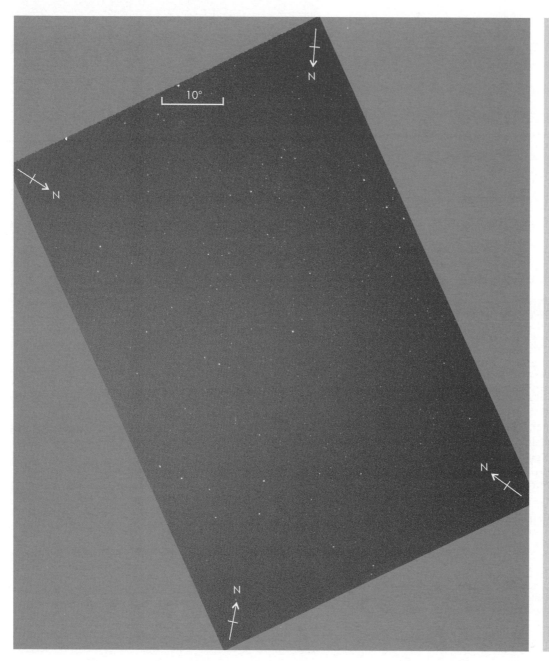

Figure 2.18.
Finding Cassiopeia –
stars visible from a
brilliant site to an
acute observer.

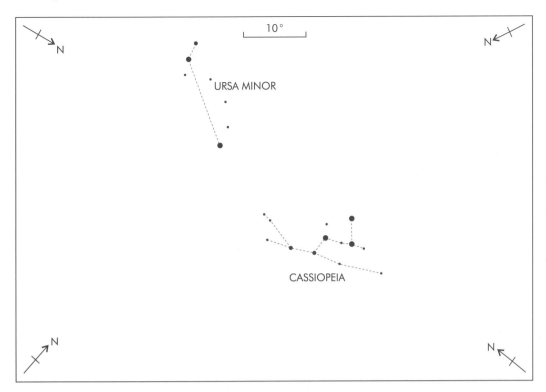

Figure 2.19. Cassiopeia showing the stars usually visible to the naked eye from a good site.

Figure 2.20. The Cassiopeia region showing the stars visible to the naked eye from a brilliant site to an acute observer (comet Hyakutake also appears, just to the left of Cassiopeia).

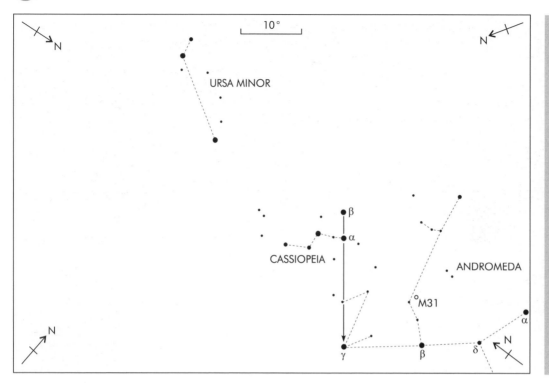

Figure 2.21. Star hopping from β Cas and γ Cas to γ And and so to the rest of Andromeda.

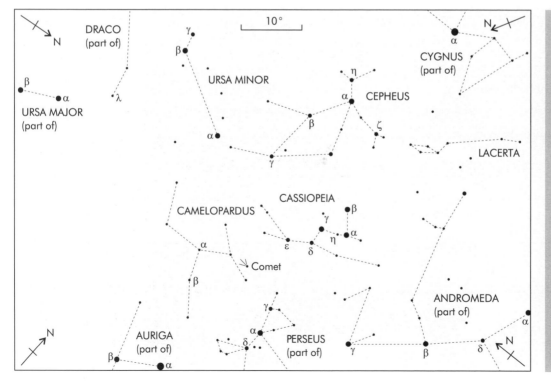

Figure 2.22. All the constellations in the Cassiopeia region.

Figure 2.23. The Perseus region – the stars visible from a good site, showing the α Per–γ And–β And–α And–β Peg asterism.

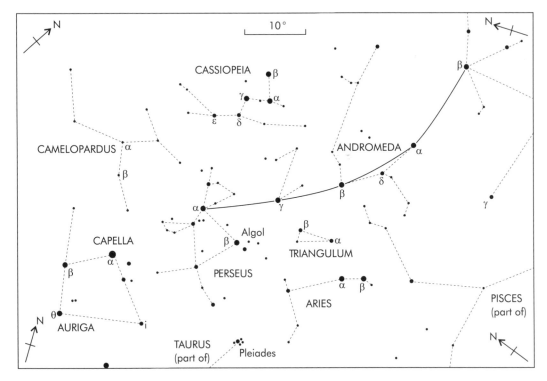

Figure 2.24. The α Per–γ And–β And–α And–β Peg asterism.

Figure 2.25.
Finding Pegasus – the
stars visible from a
typical urban, light-
polluted site.

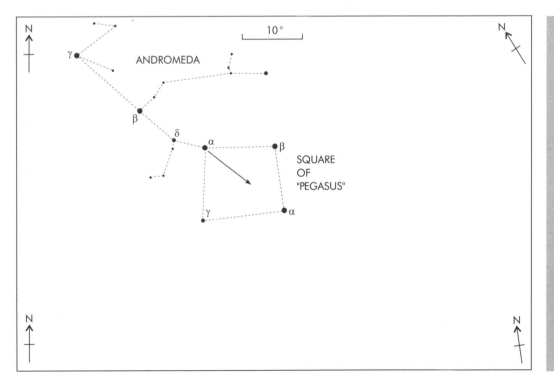

Figure 2.26. Star
hopping from
Andromeda to the
Square of Pegasus
(the square is actually
formed by α And,
α Peg, β Peg and
γ Peg).

Figure 2.27.
Finding Pegasus – the stars visible from a good site.

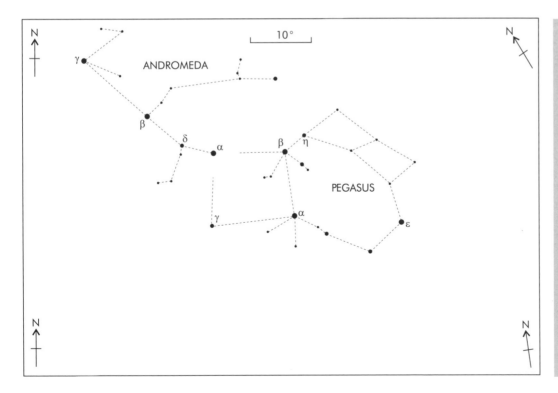

Figure 2.28.
Pegasus showing the stars usually visible to the naked eye from a good site.

Figure 2.29. Pegasus showing the stars visible to the naked eye from a brilliant site to an acute observer (Saturn is just to the left of Aquarius).

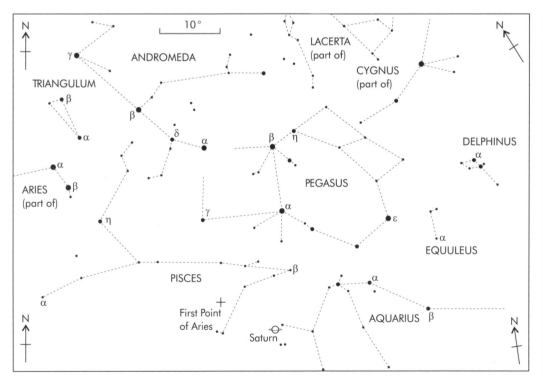

Figure 2.30. All the constellations in the Pegasus region.

Figure 2.31. Finding Cygnus – the stars visible from a typical urban, light-polluted site.

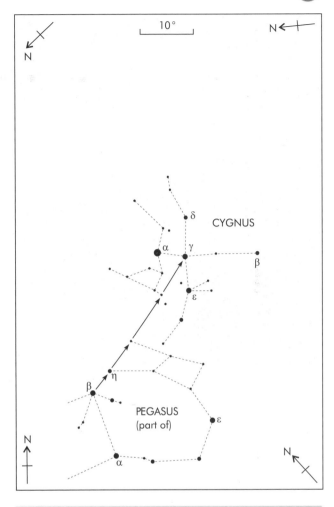

Figure 2.32. Star hopping from β and η Peg to γ Cyg, and the rest of Cygnus.

Aquila

We may also hop to the southern constellation – Aquila from Cygnus. A line from α Cyg passing between γ and ε Cyg takes us to α Aql (Altair – Fig. 2.34, *overleaf*).

The Summer Triangle

The stars Deneb, Altair and Vega (α Cyg, α Aql and α Lyr) form a large bright triangle, visible during the evenings in late summer and autumn. This asterism is sometimes known as the *summer triangle*, and it provides another easy starting point for constellation recognition (Fig. 2.35, *overleaf*).

Ophiuchus

Moving onwards brings us to a region of rather faint constellations. A star hop in a curving line from β Aql through α and γ Aql leads to α Oph (Figs 2.38, 2.39 and

Figure 2.33. Finding Cygnus – the stars visible from a good site.

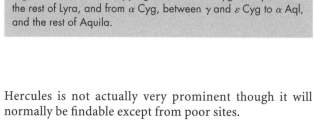

Figure 2.34. Star hopping from α and δ Cyg to α Lyr, and the rest of Lyra, and from α Cyg, between γ and ε Cyg to α Aql, and the rest of Aquila.

2.40). The rest of the equator-straddling constellation of Ophiuchus is straggling and faint but can be found on good nights.

Hercules

A slightly shallower curve from β Aql through α and γ Aql leads to α and β Her (Figs 2.38, 2.39 and 2.41). While a well known constellation from its mythological associations, Hercules is not actually very prominent though it will normally be findable except from poor sites.

Virgo

A star hop from δ through β Her in a slightly curving line brings us to Spica (α Vir) and the rest of the otherwise rather faint constellation of Virgo (Figs 2.44 to 2.48). We are however by now moving back to areas

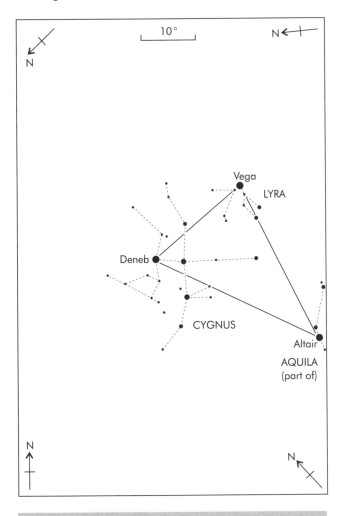

Figure 2.35. The Summer Triangle.

the constellation bears some relationship to its name (Fig. 2.51).

Canis Minor

From Leo, we may then star hop from β Leo through α Leo to α CMi (Procyon) and the rest of the small constellation of Canis Minor (Figs 2.52 and 2.53).

Auriga

We may also star hop from Ursa Major to Auriga. A line from γ UMa bisecting the line between α and β UMa (Figs 2.56 and 2.57) leads straight to α Aur (Capella) and the rest of Auriga. An easier pattern to identify in the sky, however, is the "Auriga" pentagon. This asterism is composed of α, β, θ and ι Aur plus β Tau, and is so prominent (Figs 2.58 and 2.59) that one wonders why it is not a constellation in its own right. Indeed, on some old maps, β Tau is shown as a part of Auriga and labelled as γ Aur. The Auriga pentagon is another good starting point for finding your way around the sky.

Taurus

From Auriga it is a straight line star hop from α Aur through ι Aur to α Tau (Aldebaran) and the rest of Taurus (Fig. 2.60). The two bright naked-eye galactic clusters of the Pleiades (M45) and the Hyades (C41) may be found in Taurus (Fig. 2.60).

Perseus

One may also hop in a slightly curving line from β through α Aur to α Per and the rest of Perseus (Fig. 2.60). A rather easier path though is to follow the line from Pegasus through Andromeda to α Per (Figs 2.23 and 2.60).

Gemini

From Auriga it is a slightly curving star hop to α and β Gem, also known as Castor and Pollux (Figs 2.63 and 2.64, see also Figs 2.61 and 2.62). Gemini is a prominent and easily identified constellation. An alternative route to

of the sky already seen. A better star hop to Spica may therefore be from β through α Boö in a straight line (Fig. 2.45).

Leo

It is better next to return to Ursa Major as a starting point rather than trying to use the faint stars in Virgo. We may then use the pointers in reverse, going from α UMa through β UMa in a slightly curving line to γ and then to α Leo (Regulus – Figs 2.49 and 2.50). The rest of Leo is easy to spot, since for once the shape of

Figure 2.36. The Cygnus region showing the stars visible to the naked eye from a brilliant site to an acute observer.

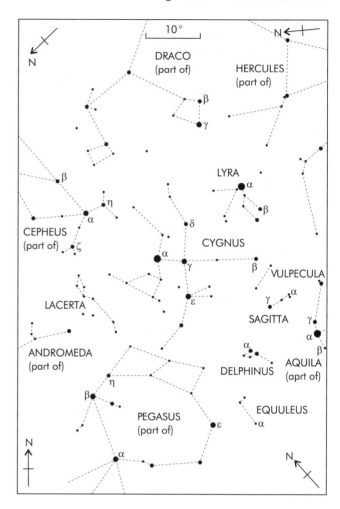

Figure 2.37. All the constellations in the Cygnus region.

Gemini is from Leo. We may go from β Leo through γ Leo in a very slightly curving line to β Gem (Figs 2.52 and 2.53).

Orion

A straight line star hop from β through γ Gem leads to α Ori (Betelgeuse) and the even more easily identified constellation of Orion (Figs 2.65 and 2.66). Orion is another good starter constellation, and strongly resembles the figure it is supposed to represent (Fig. 2.67). For observers in equatorial latitudes, for whom Ursa Major

and Crux (Section 2.3). are not circumpolar, Orion is probably the best constellation with which to commence constellation identification, although it is not visible every night.

Canis Major

From the three stars forming the belt of Orion (δ, ε and ζ Ori), our next star hop takes us to the brightest star in the sky, α CMa (Sirius). The remaining stars of Canis Major are fainter than Sirius but several are still relatively bright and all can be seen on most nights.

Figure 2.38. Stars visible in the Ophiuchus region from a typical urban light-polluted site (Jupiter is at the bottom edge, to the left of centre).

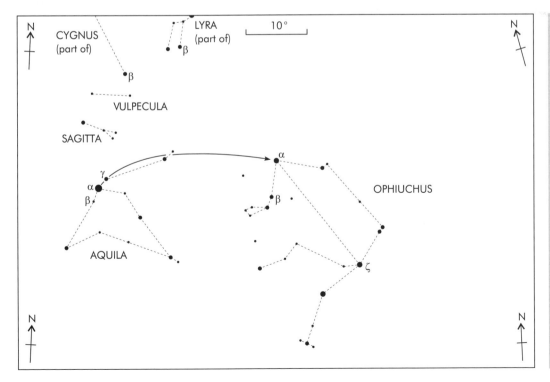

Figure 2.39. Star hopping from β Aql through α and γ Aql to α Oph and the rest of Ophiuchus.

Figure 2.40. Stars visible in the Ophiuchus region from a good site (Jupiter is at the bottom edge, to the left of centre).

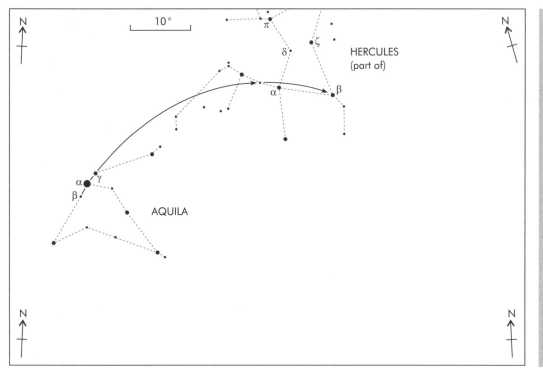

Figure 2.41. Star hopping from β Aql through α and γ Aql to α and β Her and the rest of Hercules.

Figure 2.42. The Ophiuchus region showing the stars visible to the naked eye from a brilliant site to an acute observer (Jupiter is at the bottom edge, to the left of centre).

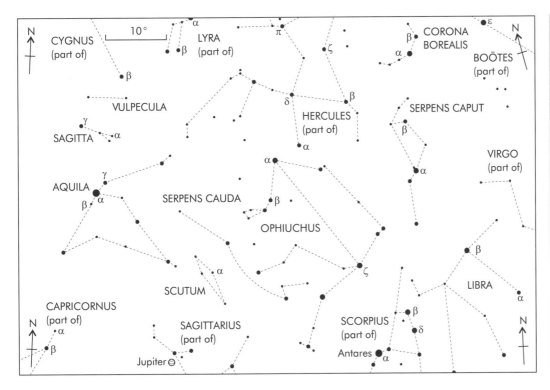

Figure 2.43. All the constellations in the Ophiuchus region.

Figure 2.44. Star hopping from δ Her through β Her to α Vir (Spica) and the rest of Virgo (or from β through α Boö) – stars visible from a typical urban light-polluted site.

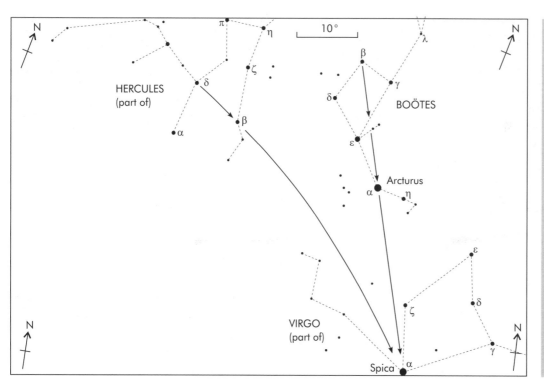

Figure 2.45. Star hopping from δ Her through β Her to α Vir (Spica) and the rest of Virgo (or from β through α Boö).

Figure 2.46. Star hopping from δ Her through β Her to α Vir (Spica) and the rest of Virgo (or from β through α Boö) – stars visible from a good site.

Figure 2.47. The Virgo region showing the stars visible to the naked eye from a brilliant site to an acute observer.

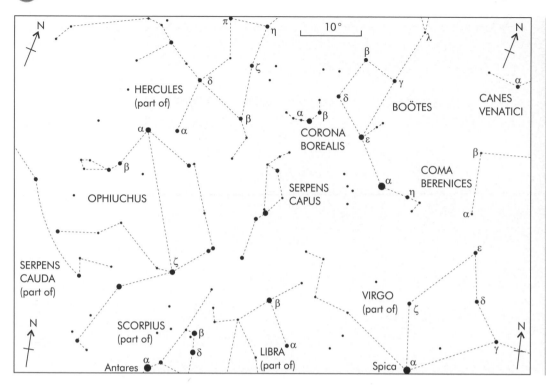

Figure 2.48. All the constellations in the Virgo region.

Figure 2.49. Leo region – the stars visible from a typical urban, light-polluted site.

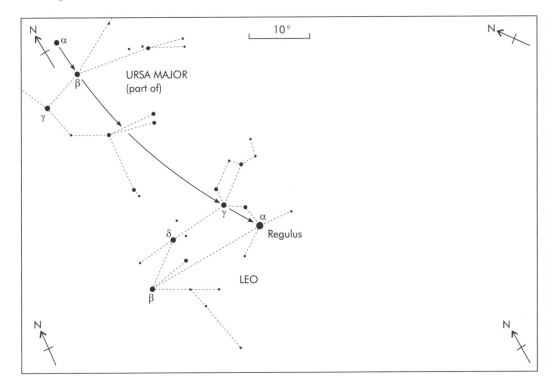

Figure 2.51. The main stars of the constellations of Leo seen as a Lion.

Figure 2.52. Leo region – the stars visible from a good site.

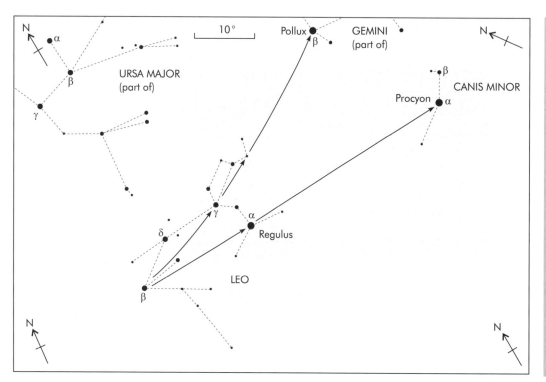

Figure 2.53. Star hopping from β through α Leo to α CMi (Procyon), and so to the rest of Canis Minor, and from β through γ Leo in a slightly curving line to β Gem (Pollux), and to the rest of Gemini.

Figure 2.54. Leo region – the stars visible from a brilliant site to an acute observer.

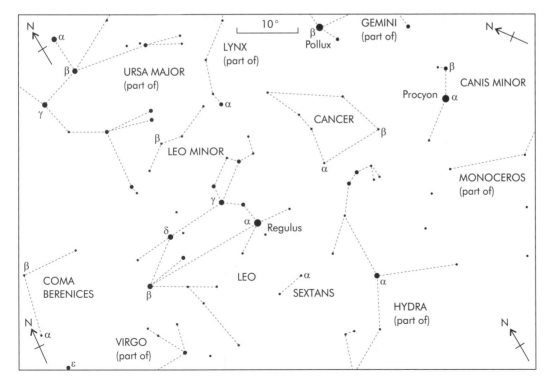

Figure 2.55. All the constellations in the Leo region.

Figure 2.56. The Auriga region – the stars visible from a typical urban, light-polluted site.

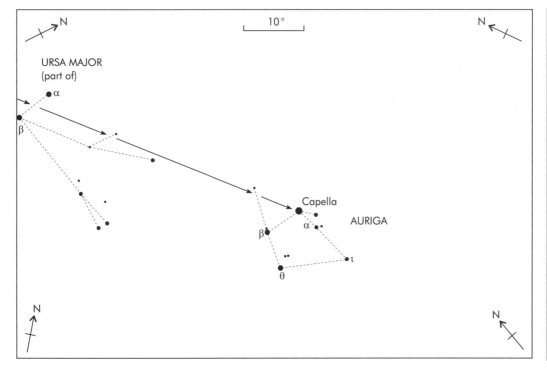

Figure 2.57. Star hopping from γ UMa and between α and β UMa to α Aur (Capella) and the rest of Auriga.

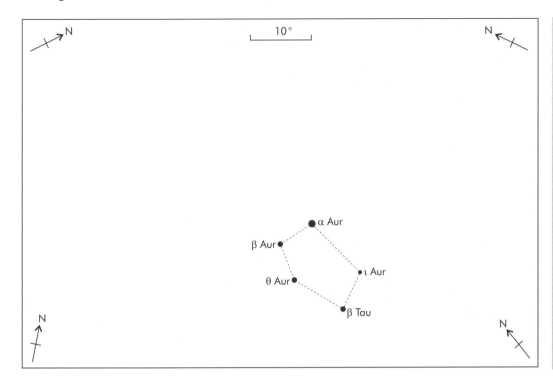

Figure 2.58. The "Auriga" pentagon – an easily recognisable asterism formed by α, β, θ, and ι Aur and by β Tau.

Figure 2.59. The Auriga region – the stars visible from a good site.

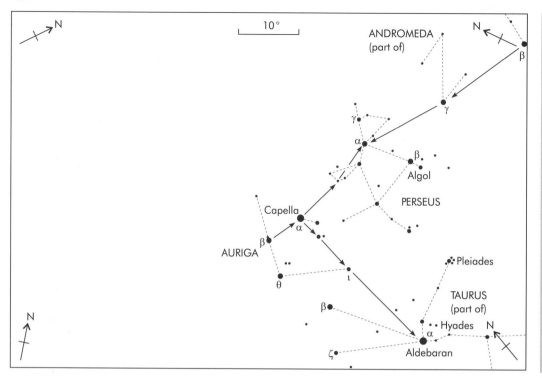

Figure 2.60. Star hopping to α Per and Perseus: (a) from β through α Aur in a slightly curving line; (b) from β through γ And in a very slightly curving line (see also Fig. 2.23) and from α through ι Aur to α Tau (Aldebaran) and the rest of Taurus.

Figure 2.61. The Auriga region – the stars visible from a brilliant site to an acute observer.

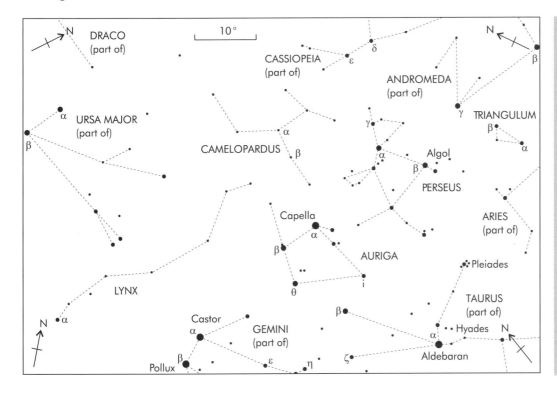

Figure 2.62. All the constellations in the Auriga region.

Figure 2.63. The Orion region – the stars visible from a typical urban, light-polluted site.

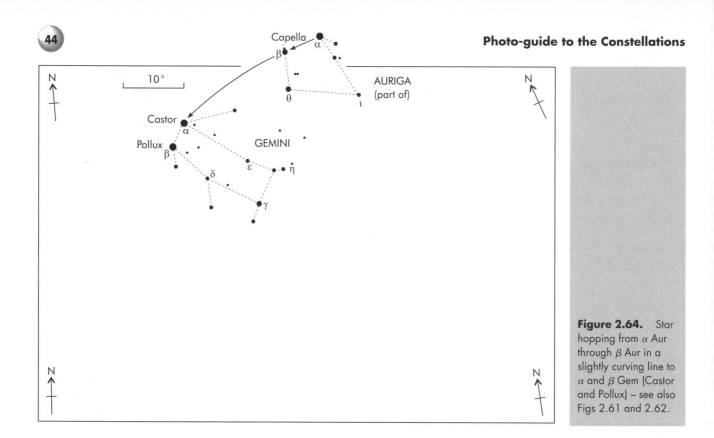

Figure 2.64. Star hopping from α Aur through β Aur in a slightly curving line to α and β Gem (Castor and Pollux) – see also Figs 2.61 and 2.62.

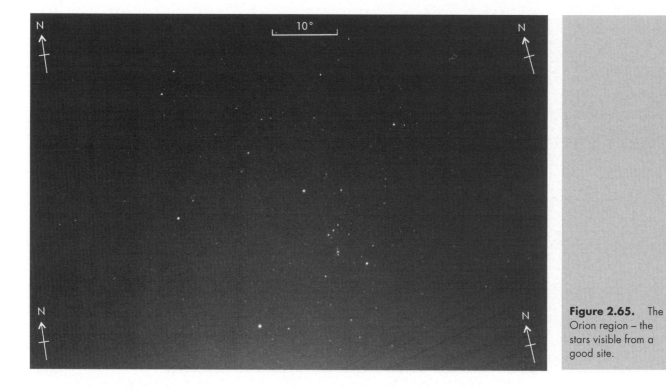

Figure 2.65. The Orion region – the stars visible from a good site.

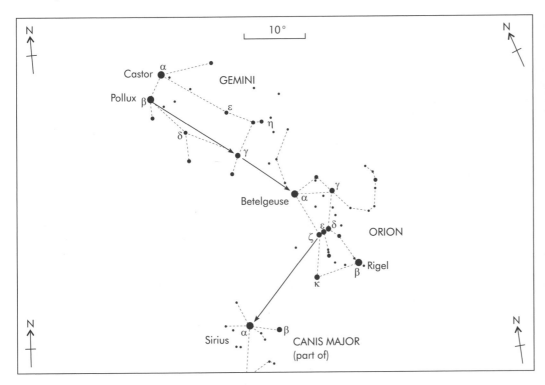

Figure 2.66. Star hopping from β through γ Gem to α Ori (Betelgeuse) and the rest of Orion. Then from Orion's belt to α CMa (Sirius) and the rest of Canis Major.

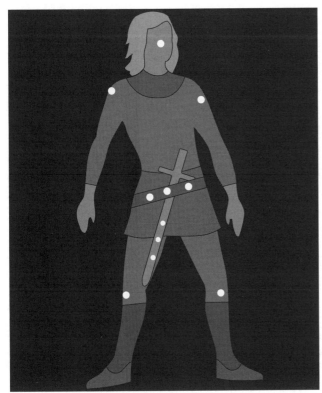

Figure 2.67. Orion as the Hunter.

Figure 2.68. The Orion region – the stars visible from a brilliant site to an acute observer.

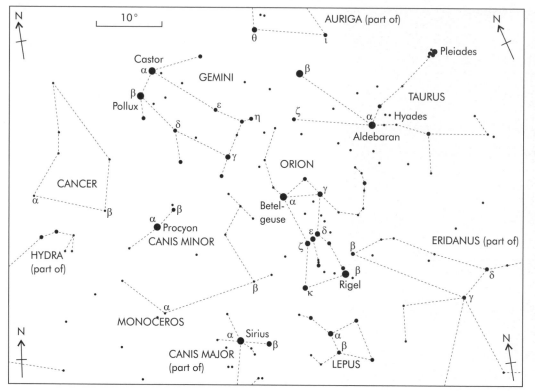

Figure 2.69. All the constellations in the Orion region.

2.2.2 The Minor Constellations

The remaining constellations for northern hemisphere observers are listed below. Their principal stars are fainter than those of the major constellations, and usually, but not always, they are quite small. Once the major constellations have been learnt it is quite easy to fill in the remainder. The same principle is used: that of star-hopping from known stars to the new constellation. With the minor constellations, this process often reduces to just finding the gap between known major constellations. The table below lists the minor constellations together with suitable major constellations as starting points for finding them. A few were included for reasons of forming useful stepping-stones during the previous section, and they are listed again here.

The Minor Constellations for Northern Hemisphere Observers

Minor Constellation	Starting point
Aquarius	Aquila, Pegasus
Aries	Andromeda, Perseus, Taurus
Camelopardus	Auriga, Cassiopeia, Perseus, Ursa Major
Cancer	Canis Minor, Gemini, Leo
Canes Venatici	Boötes, Leo, Ursa Major
Cepheus	Cassiopeia, Cygnus, Ursa Minor
Cetus	Taurus
Coma Berenices	Boötes, Leo, Ursa Major
Corona Borealis	Boötes
Delphinus	Aquila, Cygnus, Pegasus
Draco	Boötes, Cygnus, Ursa Major, Ursa Minor
Equuleus	Aquila, Pegasus
Hercules	Aquila, Boötes, Lyra
Hydra	Canis Minor, Centaurus, Leo, Scorpius
Lacerta	Andromeda, Cassiopeia, Cygnus, Pegasus
Leo Minor	Leo, Ursa Major
Lepus	Canis Major, Orion
Libra	Boötes, Centaurus, Scorpius
Lynx	Auriga, Gemini, Ursa Major
Monoceros	Canis Major, Canis Minor, Gemini, Orion
Ophiuchus	Aquila, Lyra, Scorpius
Pisces	Andromeda, Aries, Pegasus
Sagitta	Aquila, Cygnus, Lyra
Scutum	Aquila, Sagittarius
Serpens Cauda	Aquila, Sagittarius
Serpens Caput	Boötes, Scorpius
Sextans	Leo
Triangulum	Andromeda, Perseus, Taurus
Virgo	Boötes, Leo
Vulpecula	Aquila, Cygnus, Lyra

2.3 Southern Hemisphere Observers (Latitudes –20° to –90°)

Southern observers, like northern observers, can hope to see fifty or sixty constellations at some time throughout the year. But just twenty of these will be sufficient to find your way around the sky. These brighter or major constellations are listed below.

2.3.1 The Major Constellations

The major constellations for southern hemisphere observers are:

The Major Constellations for Southern Hemisphere Observers

Aquila	Ara	Canis Major	Carina
Centaurus	Cetus	Crux	Gemini
Grus	Leo	Lupus	Orion
Pavo	Pisces Austrinus	Puppis	Sagittarius
Scorpius	Taurus	Triangulum Australe	Vela

In the southern hemisphere, Crux is a good starting point for constellation identification. Its main stars are circumpolar for any latitudes south of –35°, and it is easily recognisable *ab initio*. It therefore forms the equivalent of Ursa Major as a starting point for most southern observers. Like Ursa Major, the appearance of Crux in the sky changes with its orientation, but it is a compact, bright constellation and usually easily recognisable in any orientation (Fig. 2.70).

Crux

How then do you find Crux for the first time? The procedure is similar to that used by a northern hemisphere observer starting off to find Ursa Major (Section 2.2.1). Just as with Ursa Major, therefore, the appearance of the constellation will change depending upon circumstances. From an urban observing site probably only the three brightest stars of the constellation will be visible (Fig. 2.71). In poor, hazy conditions and with a bright Moon, you may not be able to see any stars at all. From a reasonable site, on a good clear night, without a Moon, on the other hand, you will probably be able to see eight or ten stars in the constellation (Fig. 2.72). The appearance will also depend upon the orientation of the constellation (Fig. 2.70), but the symmetrical and compact nature of Crux makes this less significant than for the rambling Ursa Major. Finally, the constellation may be at different altitudes above the

horizon, or from some latitudes, some or all the stars may be below the horizon.

With these problems in mind, therefore, it is best to select a reasonably clear moonless night, and if need be to move away from town and other bright lights, for your first attempt at constellation identification. As you are able to recognise more and more constellations, you will

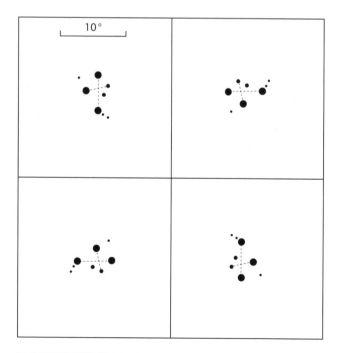

Figure 2.70. The changing appearance of Crux at different orientations.

Figure 2.71. The Centaurus region – the stars visible from a typical urban, light-polluted site.

Figure 2.72. The Centaurus region – the stars visible from a good site.

Having chosen a reasonable night, arrived at a reasonable observing site, and oriented yourself to face south, what then? First get an idea of the area of sky over which you will need to search. Crux is about 30° from the South Pole. You will therefore need to search over a circle extending from 30° either side of south, and from near the horizon to near the zenith. This radius is about four times the width of the clenched fist at arm's length (Fig. 1.1). The constellation's four main stars extend over a distance of about 8° (the clenched fist at arm's length). Step these two distances out so that you can visualise the scale of the constellation and of the search area. Then with the possible appearances in mind (Fig. 2.70) look among the brighter stars in the search area until you can find the cross-shaped pattern of Crux. You should be able to spot the constellation within a minute or two.

If after ten or fifteen minutes' searching you have not found the constellation, then you are probably not facing south, so check your orientation. Alternatively, if your latitude is north of about –35°, then some of the main stars of Crux may be below your horizon (see Chapter 4, also astronomy magazines and some national newspapers publish monthly sky charts showing the stars visible for the appropriate time of year, so you can check these if you suspect that Crux is below the horizon). In the latter case you can either try again at a different time of night or later in the year, or start with a different constellation (see below). An alternative starter constellation, Carina (see below), also provides pointers to the Southern Cross (Figs 2.71, 2.72 and Fig. 2.73, *overleaf*) which may provide means of getting going. If Crux should be visible, and you are oriented correctly to see it, and yet you still cannot find it, then you will probably have to find someone who can point it out to you directly.

Assuming that you have been successful in finding the main stars of Crux, you can then start to become familiar with their individual names. But unlike Ursa Major only two of the stars have proper names. The southern-most star of the cross is Acrux, and the northern-most is Gacrux. All the main stars though do have designations under the Bayer system as α Crucis (usually abbreviated as α Cru – see Section 1.10 for the genitives and abbreviations of all the constellation names), β Cru, γ Cru and δ Cru (Fig. 2.73).

Once you are familiar with the main stars of Crux, you should start looking for the fainter ones, though since this is a small constellation there are relatively few more stars to find (Figs 2.4, 2.9 and 2.12).

find that it is not essential to have good observing conditions in order to find your way around the sky, but it will help considerably to begin with. You will also need to know the approximate direction of the south on the ground. If you do not know this already for your area, then it can easily be obtained from a good-quality map.

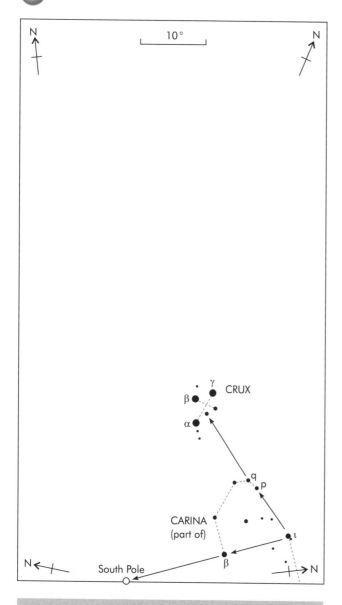

Figure 2.73. Star hopping from ι Car through p and q Car to Crux. ι Car and β Car form pointers to the Southern Pole, although there is no bright star near that point to identify it in the sky.

Figure 2.74. The Centaurus region – the stars visible from a brilliant site to an acute observer.

Again like Ursa Major (Fig. 2.9), Crux is also used to delimit a specific area of the sky. But for Crux this is a simple box hardly larger than the area covered by the main stars of the constellation.

Centaurus

Once you can identify Crux (or any other "starter" constellation), finding subsequent constellations is straightforward. Pairs of stars from the known constellation are used to point to the next (star hopping – Section 1.4). Crux provides a short star hop from β Cru to β Cen and then on to α Cen and the rest of Centaurus (Figs 2.71, 2.72 and 2.75). The brightest of the globular clusters, ω Cen (C80, NGC 5139) is easily seen on a reasonable night on the line between θ Cen and γ Cru (Figs 2.74 and 2.76).

Triangulum Australe

A slightly longer hop from γ Cru through β Cru leads to α TrA and the relatively small number of other stars making up the constellation of Triangulum Australe (Figs 2.74, 2.75 and 2.76).

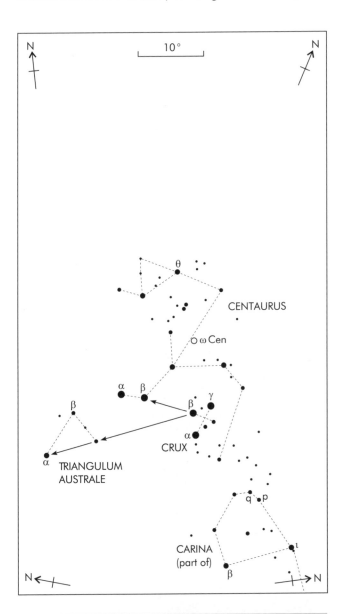

Figure 2.75. Star hopping from Crux to β and α Cen and the rest of Centaurus, and from γ Cru through β Cru to α TrA and the rest of Triangulum Australe.

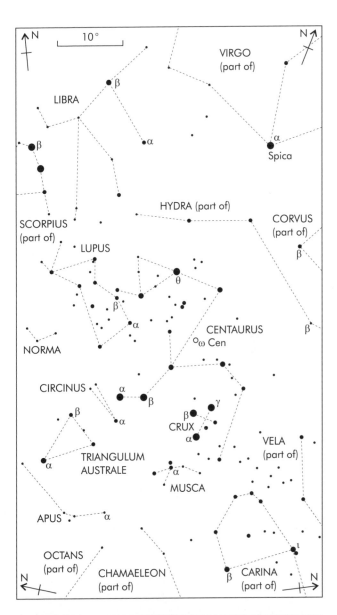

Figure 2.76. All the constellations in the Centaurus region.

Grus and Pavo

Centaurus and Triangulum Australe together provide the next star hop: from α Cen through α TrA to α Pav and the rest of Pavo in a slightly curving line, and from the same two stars but in a straight line to β Gru and the rest of Grus (Figs 2.77, 2.78 and Fig. 2.80, *overleaf*).

Scorpius

The prominent constellation of Scorpius is found by hopping from β Cen through α Cen to θ Sco in a slightly curving line (Figs 2.77, 2.79 and Fig. 2.80, *overleaf* and see also Figs 2.38, 2.39 and 2.40). It is perhaps more easily found though by hopping to Antares (α Sco) on a line from α Oph and which then goes between η and ζ Oph (Figs 2.88, 2.90 and 2.91) or from Hercules (Fig. 2.42 and Fig. 2.83).

Ara and Lupus

The moderately faint constellations of Ara and Lupus can easily be found between Scorpius and Triangulum Australe (Ara) and between Scorpius and Centaurus (Lupus – Figs 2.81 and 2.82)

Sagittarius

A continuation of the line from α Cen to θ Sco then leads on to the main part of Sagittarius (Figs 2.77, 2.79 and Fig. 2.80, *overleaf*).

Phoenix and Tucana

From Grus we may progress in a star hop from α Gru through β Gru to α Phe and the rest of the rather inconspicuous constellation of Phoenix (Figs 2.84, 2.85 and 2.86). The second brightest globular cluster, 47 Tuc (C106, NGC 104), is to be found on a line between α TrA and α Eri (Achernar). The minor constellation of Tucana is then next to Phoenix (Figs 2.87 and 2.88) and the Small Magellanic Cloud (SMC or Nubecular Minor) is to be found there (Fig. 2.82).

Aquila and Capricornus

The rather faint constellation of Capricornus can be found through star hops to α Cap and β Cap from ε Sgr through ζ Sgr in a straight line, or from Altair (α Aq1 – Figs 2.34 and 2.35) through θ Aql in a slightly curving line (Figs 2.89, 2.90 and 2.92).

Figure 2.77. The Pavo region – the stars visible from a typical urban, light-polluted site (Jupiter appears at the top, to the right of centre).

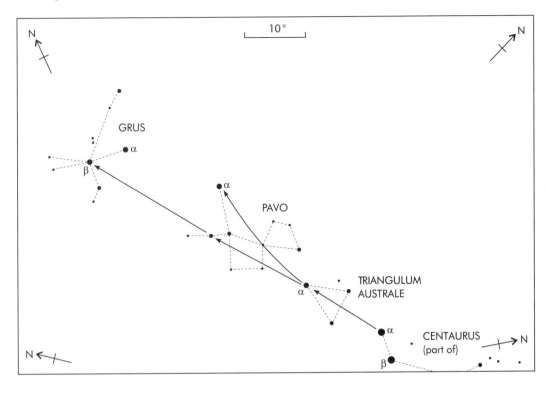

Figure 2.78. Star hopping from α Cen through α TrA to: (a) α Pav and the rest of Pavo in a slightly curving line and (b) β Gru and the rest of Grus in a straight line.

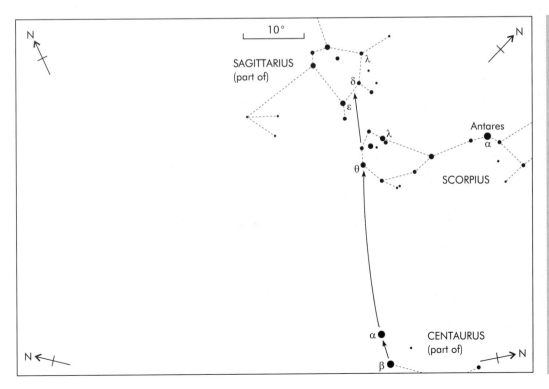

Figure 2.79. Star hopping from β Cen through α Cen to θ Sco and the rest of Scorpius in a slightly curving line, then continuing on to Sagittarius.

Figure 2.80. The Pavo region – the stars visible from a good site (Jupiter appears at the top, to the right of centre).

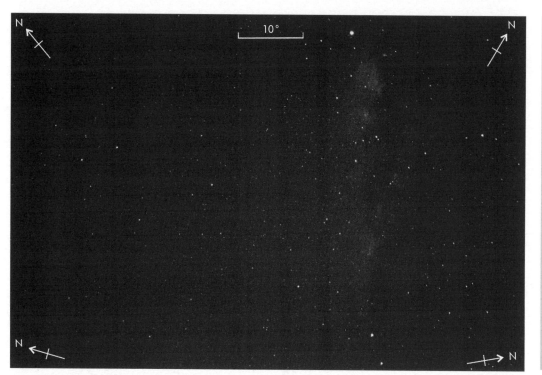

Figure 2.81. The Pavo region – the stars visible from a brilliant site to an acute observer (Jupiter appears at the top, to the right of centre).

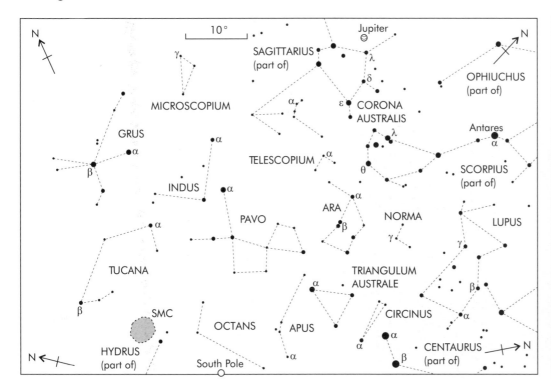

Figure 2.82. All the constellations in the Pavo region.

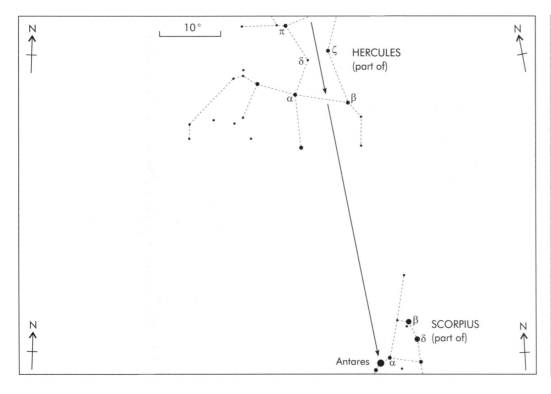

Figure 2.83. Star hopping from the centre line of Hercules to α Sco (Antares) and the rest of Scorpius (see main photograph – Fig. 2.42).

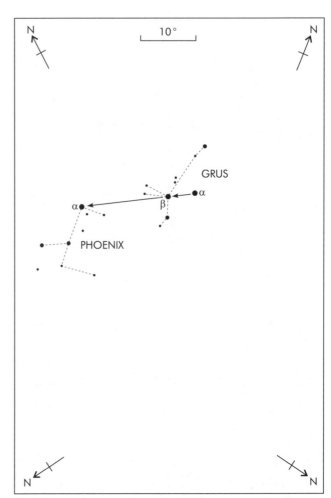

Figure 2.84. The Tucana region – the stars visible from a typical urban, light-polluted site.

Figure 2.85. Star hopping from α Gru through β Gru to α Phe and the rest of Phoenix.

Figure 2.86. The Tucana region – the stars visible from a good site.

Figure 2.87. The Tucana region – the stars visible from a brilliant site to an acute observer.

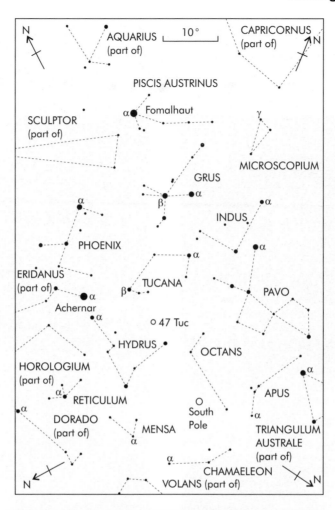

Figure 2.88.   All the constellations in the Tucana region.

Carina and Orion

Returning now to Crux, there is a short star hop from
β Cru between α Cru and δ Cru to q and p Car and the rest
of Carina (Fig 2.73). Alternatively from Orion (Figs 2.66
and 2.68) we can star hop to Canopus (α Car – Figs 2.95,
2.96 and 2.97). Carina at one time formed a part of the
constellation of Argo, but this was divided during the eighteenth century into Carina, Puppis and Vela. The existence
of this precursor explains the otherwise rather puzzling
meaning of Carina which is the Keel. Carina provides
pointers to the South Pole (Figs 2.71 and 2.73). A straight
line star hop from ι Car through β Car by about twice their

separation leads directly to the pole. However there is no
convenient bright star to mark the Southern Pole in the
way that Polaris (α UMi) marks the Northern Pole, so its
position as given by the Carina pointers must be imagined.

Canis Major

The star hop from Orion to Carina passes Sirius (α CMa),
the brightest star in the night sky, and the rest of the constellation of Canis Major (Fig. 2.96). Sirius is so bright that
identifying it is not normally a problem, but if needed, the
stars of Orion's belt point directly towards it (Fig. 2.66).

Figure 2.89. The Sagittarius region – the stars visible from a typical urban, light-polluted site (Jupiter appears just below centre).

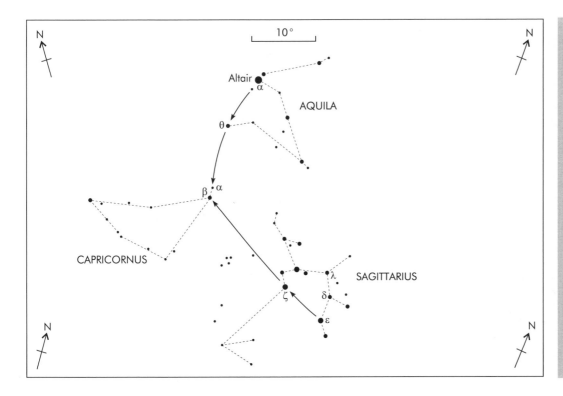

Figure 2.90. Star hopping from ε Sgr through ζ Sgr in a straight line, or from Altair (α Aql) through θ Aql in a slightly curving line to α Cap and β Cap and the rest of Capricornus.

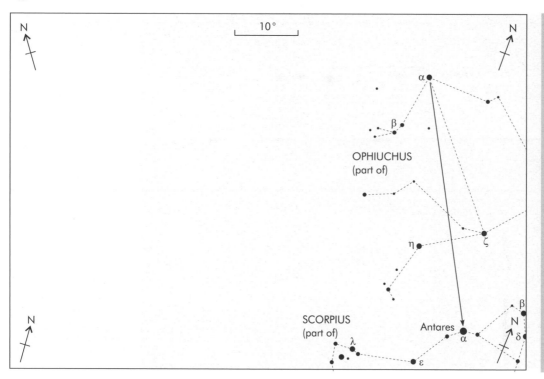

Figure 2.91. An alternative star hop to Scorpius – from α Oph and between η and ζ Oph to α Sco (Antares).

Figure 2.92. The Sagittarius region – the stars visible from a good site (Jupiter appears just below centre).

Figure 2.93. The Sagittarius region – the stars visible from a brilliant site to an acute observer (Jupiter appears just below centre).

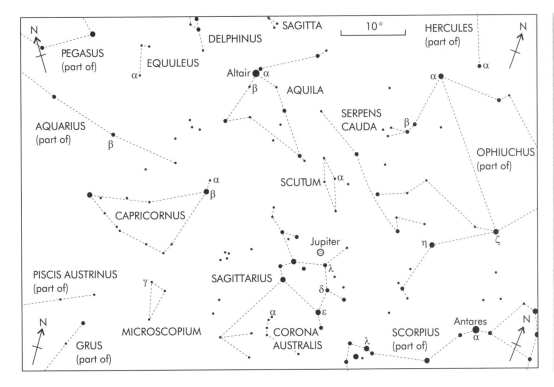

Figure 2.94. All the constellations in the Sagittarius region.

Figure 2.95. The Carina region – the stars visible from a typical urban, light-polluted site.

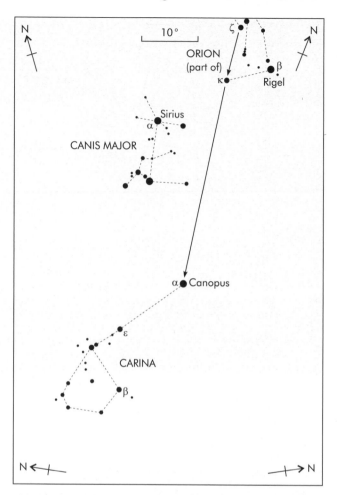

Figure 2.96. Star hopping from ζ Ori through κ Ori to α Car (Canopus) and the rest of Carina.

Puppis and Vela

From Sirius (α CMa) we may hop through ε CMa down to Puppis and on to the bright Wolf–Rayet star, γ Vel, in a slightly curving line (Fig. 2.98).

Dorado and Mensa

The minor constellations of Dorado and Mensa are adjacent to Carina (Fig 2.100, *overleaf*). Although they are inconspicuous their border contains the spectacular Large Magellanic Cloud (LMC or Nubecular Major – Fig. 2.100, *overleaf*) with its numerous nebulae, and which is the nearest external galaxy to us.

Gemini and Taurus

Star hopping from β Ori (Rigel) through α Ori (Betelgeuse) brings us to γ Gem and the rest of Gemini (Figs 2.66 and 2.69, see also Fig. 2.64). Taurus is next to Gemini and Orion, and a short curving star hop from κ Ori through γ Ori leads to Aldebaran (α Tau – Fig. 2.69 and

Figure 2.97. The Carina region – the stars visible from a good site.

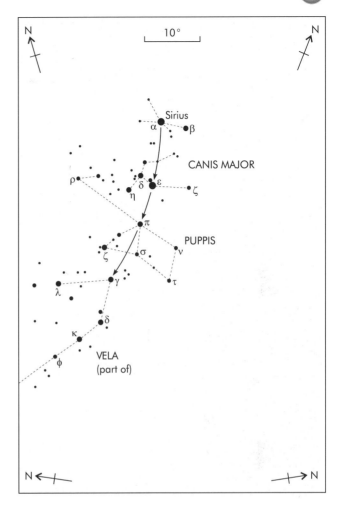

Figure 2.98. Star hopping from α CMa (Sirius) in a slightly curving line through ε CMa to π Pup and the rest of Puppis and then onwards to γ Vel.

Fig. 2.101, *overleaf*, see also Figs 2.60 and 2.61) and the rest of Taurus. This latter constellation includes the two bright galactic star clusters of the Pleiades and Hyades (Fig. 2.69).

Cetus

From either Orion (using κ and β Ori) or from Taurus (using α and λ Tau) it is a straight line star hop to α Cet and the rest of Cetus (Figs 2.102, 2.103 and 2.104). The star Mira, on the line between α and β Cet (Fig. 2.106), is the prototype for all long-period variable stars. However it

can range in magnitude from $+2^m$ to $+9^m$ and so may not always be visible to the naked eye.

Pisces Austrinus

Fomalhaut (α PsA) may be found from either of two longish star hops: firstly from α Cet through β Cet or secondly from β And through γ Peg (see Fig. 2.21 for Andromeda), the latter in a slightly curving line (Figs 2.107, 2.108 and 2.109). The remaining stars in Pisces Austrinus are rather faint, but can be seen on a good night.

Figure 2.99. The Carina region – the stars visible from a brilliant site to an acute observer.

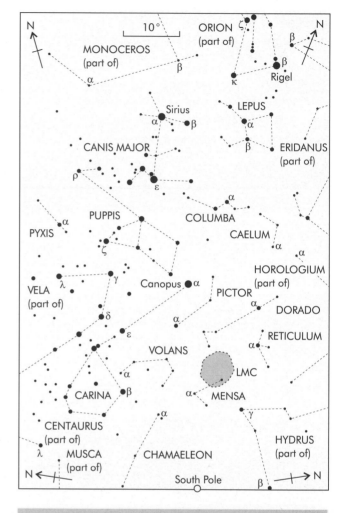

Figure 2.100. All the constellations in the Carina region.

Aquarius

α And and α Peg give a straight line star hop to α Aqr and the rest of Aquarius (Fig. 2.108).

Hydra, Leo and Vela

Returning to Carina, a very short star hop from β Car through ι Car leads to κ Vel and the rest of Vela (Figs 2.112,

2.113 and 2.114; see also Fig. 2.98). Carrying on over a much longer hop then leads to α Hya, and the otherwise faint and straggling constellation of Hydra (Figs 2.112, 2.113 and 2.114). Other star hops to α Hya are from β through α CMa (Sirius – Figs 2.112, 2.114 and 2.115), and from γ through α Leo (Regulus – Figs 2.118, 2.119 and 2.120). Leo is a good starter constellation because of its resemblance to a lion (Fig. 2.51). It can be found from Ursa Major (Fig. 2.50) or by looking about 60° to the west of α Ori (Betelgeuse – about eight times the width of the clenched fist at arm's length).

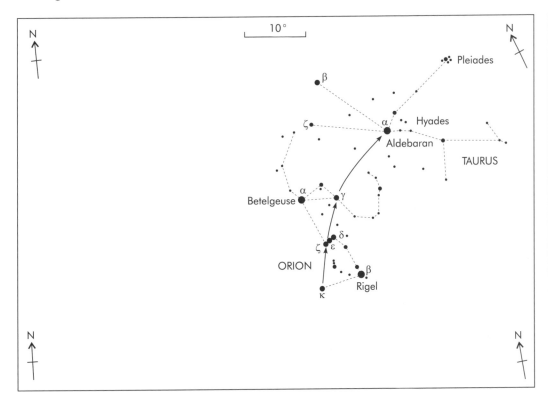

Figure 2.101.
Star hopping from
κ Ori through γ Ori to
Aldebaran (α Tau)
and the rest of Taurus
(see main photograph
– Fig. 2.68).

Figure 2.102.
The Cetus region –
the stars visible from
a typical urban, light-
polluted site (Saturn is
the bright object in
the top right-hand
corner).

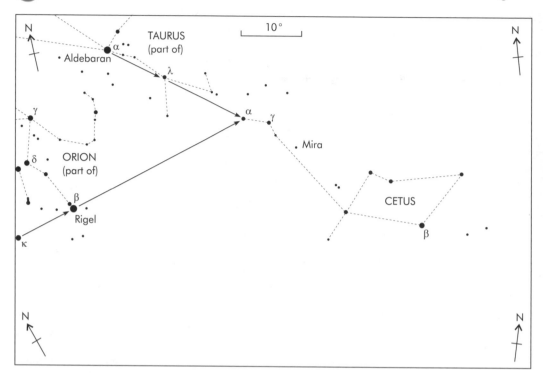

Figure 2.103.
Star hopping from
κ Ori through β Ori
(Rigel), or from α Tau
(Aldebaran) through
λ Tau to α Cet and
the rest of Cetus.

Figure 2.104.
The Cetus region –
the stars visible from
a good site (Saturn is
the bright object in
the top right-hand
corner).

Figure 2.105. The Cetus region – the stars visible from a brilliant site to an acute observer (Saturn is the bright object in the top right-hand corner).

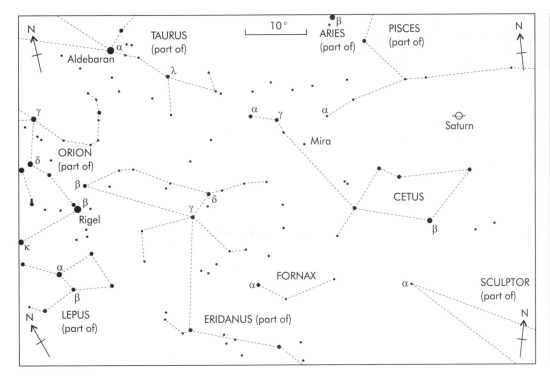

Figure 2.106. All the constellations in the Cetus region.

Figure 2.107. The Pisces region – the stars visible from a typical urban, light-polluted site (Saturn is just above the centre).

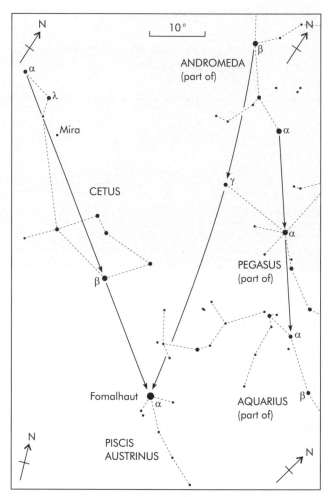

Figure 2.108. Star hopping: (a) in a straight line from α Cet through β Cet, or in a slightly curving line from β And through γ Peg, to α PsA (Fomalhaut) and the rest of Pisces Austrinus and (b) in a straight line from α And through α Peg to α Aqr and the rest of Aquarius.

Figure 2.109. The Pisces region – the stars visible from a good site (Saturn is just above the centre).

Figure 2.110. The Pisces region – the stars visible from a brilliant site to an acute observer (Saturn is just above the centre).

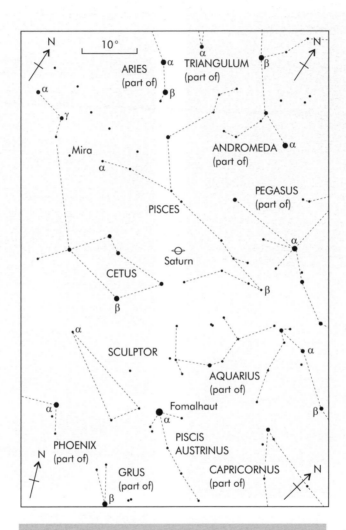

Figure 2.111.   All the constellations in the Pisces region.

Figure 2.112.
The Vela region – the stars visible from a typical urban, light-polluted site.

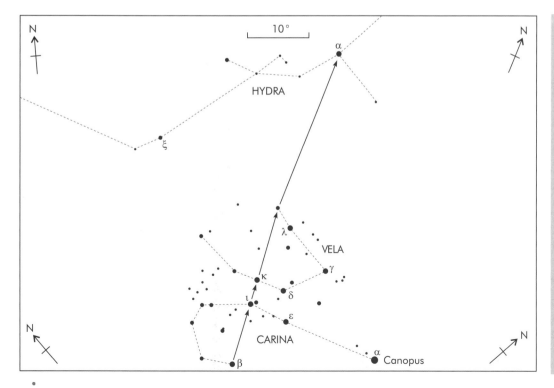

Figure 2.113.
Star hopping from β Car through ι Car to κ Vel and the rest of Vela – then onwards to α Hya.

Figure 2.114.
The Vela region – the stars visible from a good site.

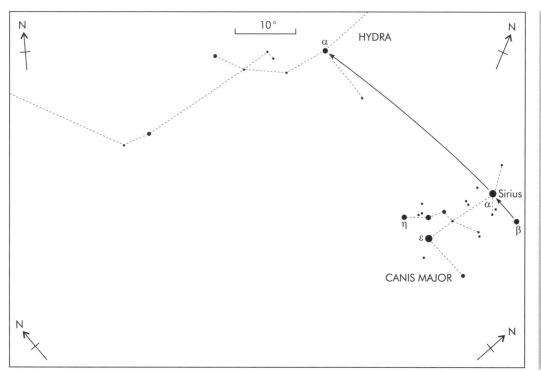

Figure 2.115.
Star hopping from β CMa through α CMa (Sirius) to α Hya.

Figure 2.116.
The Vela region – the stars visible from a brilliant site to an acute observer.

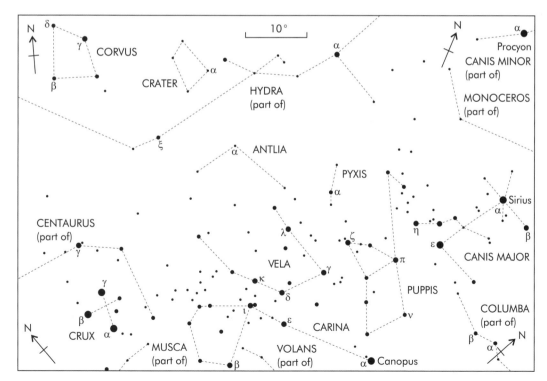

Figure 2.117. All the constellations in the Vela region.

Figure 2.118.
The Hydra region – the stars visible from a typical urban, light-polluted site (Mars appears just above the centre).

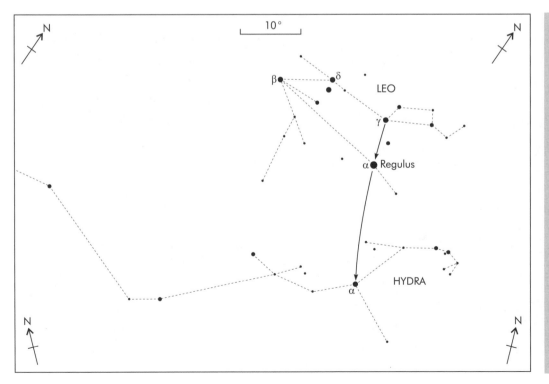

Figure 2.119.
Star hopping from γ Leo through α Leo (Regulus) to α Hya and the rest of Hydra.

Figure 2.120.
The Hydra region – the stars visible from a good site (Mars appears just above the centre).

Figure 2.121.
The Hydra region – the stars visible from a brilliant site to an acute observer (Mars appears just above the centre).

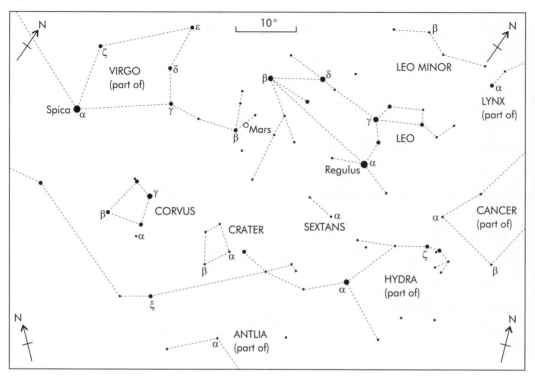

Figure 2.122. All the constellations in the Hydra region.

2.3.2 The Minor Constellations

The remaining constellations for southern hemisphere observers are listed below. Their principal stars are fainter than those of the major constellations, and usually, but not always, they are quite small. Once the major constellations have been learnt it is quite easy to fill in the remainder. The same principle is used: that of star-hopping from known stars to the new constellation. With the minor constellations, this process often reduces to just finding the gap between known major constellations. The table below lists the minor constellations together with suitable major constellations as starting points for finding them. A few were included for reasons of forming useful stepping-stones during the previous section, and they are listed again here.

The Minor Constellations for Southern Hemisphere Observers

Minor Constellation	Starting point
Antlia	Centaurus, Puppis, Vela
Apus	Triangulum Australe

Aquarius	Cetus, Pisces Austrinus
Caelum	Canis Major, Carina
Cancer	Gemini, Leo
Canis Minor	Canis Major, Gemini
Capricornus	Aquila, Pisces Austrinus, Sagittarius
Chamaeleon	Carina, Crux
Circinus	Centaurus, Triangulum Australe
Columba	Canis Major
Corona Australis	Sagittarius, Scorpius
Corvus	Centaurus
Crater	Centaurus, Vela
Dorado	Carina
Eridanus	Cetus, Orion
Fornax	Cetus
Horologium	Carina, Grus
Hydra	Centaurus, Leo
Hydrus	Pavo
Indus	Grus, Pavo, Sagittarius
Lepus	Canis Major, Orion
Libra	Centaurus, Lupus, Scorpius

Mensa	Carina
Microscopium	Grus, Sagittarius
Monoceros	Canis Major, Orion
Musca	Carina, Centaurus, Crux
Norma	Ara, Lupus, Scorpius, Triangulum Australe
Octans	Pavo, Triangulum Australe
Ophiuchus	Sagittarius, Scorpius
Phoenix	Grus, Pisces Austrinus
Pictor	Carina
Pyxis	Puppis, Vela
Reticulum	Carina
Sculptor	Cetus, Pisces Austrinus
Scutum	Aquila, Sagittarius
Serpens Cauda	Aquila, Sagittarius
Sextans	Leo
Telescopium	Ara, Pavo, Sagittarius
Tucana	Grus, Pavo
Virgo	Leo
Volans	Carina

Canis Minor	Cetus	Gemini	Leo
Orion	Pegasus	Sagittarius	Scorpius
Taurus			

There is no single best starter constellation for equatorial observers, because all constellations are below the horizon for some of the time (Chapter 4). If visible, Ursa Major or Crux may be used (Sections 2.2.1 and 2.3.1). Otherwise the following are probably the best constellations to start with if observing before midnight, when they will be high in the sky during the indicated times of year:

Starter Constellations for Equatorial Observers	
Time of year	Starter Constellation(s)
January–February	Canis Major or Orion
February–May	Leo
May–June	Scorpius
July–August	Aquila or Scorpius
September–October	Aquila or Pegasus
November–December	Orion or Pegasus

2.4 Equatorial Observers (Latitudes 30° to –30°)

From the equator itself, all stars and constellations in the sky are visible at one time or another (see Chapter 4). The opposing polar regions will however become increasingly hidden for observers to the north or south of the equator. In practice, observers in the above range of latitudes will be able to use most of the information in sections 2.2 and 2.3 as well as that in this section.

2.4.1 The Major Constellations

The major constellations for equatorial observers are listed below. These are the ones that will pass overhead, or nearly so. Most of the other major constellations listed for northern and southern observers will however also be usable by observers in equatorial latitudes.

The Major Constellations for Equatorial Observers			
Andromeda	Aquila	Boötes	Canis Major

Finding your first constellation is probably more difficult for an equatorial observer than for northern or southern observers, because there is no single part of the sky guaranteed to contain the constellation being sought. The starter constellations listed above however will be found somewhere in a broad band stretching from east to west through the zenith, and if you are observing before midnight at the indicated times of year, they will be well above the horizon (see also Chapter 4). Help from someone who already knows the constellations may well be advantageous. The starter constellations can be found from the following images and diagrams:

Constellation	Figure*
Aquila	2.43, 2.94
Canis Major	2.100, 2.117
Leo	2.55, 2.122
Orion	2.69
Pegasus	2.30
Scorpius	2.82

* The figure numbers refer to the diagrams showing all the constellations contained in an image. The photographs showing appearances of that area of sky under various observing conditions, and star-hopping routes will be found immediately prior to these figures. Additional figures containing parts of the constellations are listed in Section 1.10.

Just as for northern and southern observers, once you can recognise one constellation, that can then be used to star hop to the next. The remaining major constellations for equatorial observers are to be found on the following figures, and routes to them are given in Sections 2.2.1 and 2.3.1:

Constellation	Figure*
Andromeda	2.24, 2.30
Boötes	2.13
Canis Minor	2.55, 2.69
Cetus	2.106, 2.111
Gemini	2.69
Sagittarius	2.94
Taurus	2.69

* The figure numbers refer to the diagrams showing all the constellations contained in an image. The photographs showing appearances of that area of sky under various observing conditions, and star-hopping routes will be found immediately prior to these figures. Additional figures containing parts of the constellations are listed in Section 1.10.

2.4.2 The Minor Constellations

The minor constellations may be found once the major ones can be recognised, and these are listed in Sections 2.2.2 and 2.3.2.

Chapter 3

The Individual Constellations

3.1 Introduction

In Section 3.2, further details of the individual constellations and of the objects to be found within them are given. The information provided extends in many cases beyond what may be seen with the naked eye, to include details that will require the use of binoculars or a small telescope. This is so that those readers who have had their interest in astronomy pricked by learning the constellations can make further progress.

The figure numbers refer to the diagrams showing all the constellations contained in an image. Figures containing all or nearly all of the main stars of a constellation are underlined. In a few cases two images are required to encompass the whole constellation, and then those two images are bracketed and underlined. The photographs showing appearances of that area of sky under various observing conditions, and star-hopping routes, will be found immediately prior to the figures listed.

The meaning of or the legend associated with each constellation is briefly summarised. In many cases, though, it is not certain that the legend has given rise to the constellation, or there may be alternative explanations. The majority (48) of the constellations date back to ancient Greek or earlier times. The remainder were introduced as new star maps, and catalogues were produced in the sixteenth, seventeenth and eighteenth centuries. Many of the later constellations are small and faint and of relatively little significance. They may also have, to present-day opinions, what seem to be rather strange names. Thus we would hardly be likely today to name a new constellation for an air pump (Antlia) or a chemical furnace (Fornax). But we now have to live with those names.

The details given for the stars and other objects include their brightnesses (magnitudes), position (Right Ascension and Declination) and nature (such as type of variable star, globular cluster etc.). The meaning of these items is explained below. Some stars have individual names (see Section 1.7), and these are listed for each constellation. Some stars even have several names, or alternative spellings. Where one name is in much more common use than the others, it is underlined.

For nebulae, star clusters and galaxies, the Messier, Caldwell, NGC (New General Catalogue) and/or IC (Index Catalogue) numbers are given. Often these are used as the names of the objects. In all cases, knowing these identifiers will enable the objects to be found quickly in other sources, should further information be wanted.

3.1.1 Magnitudes

The brightness of stars is measured by astronomers in MAGNITUDES. This is a somewhat odd system. Firstly because it works the "wrong" way – the brighter the star, the smaller the value of its magnitude. Secondly it is a geometrical scale – that is, a difference of one magnitude corresponds to a constant multiple (unlike the normal arithmetical scales where a difference of one unit corresponds to a constant addition). Thus a star of magnitude 3 is about two and a half times brighter than one of magnitude 4, 6.3 (= 2.5^2) times brighter than one of magnitude 5, and 16 (= 2.5^3) times brighter than one of magnitude 6.

The magnitudes of many stars are listed in the discussions on each constellation in Section 3.2. To give an idea how the scale operates in practice, Orion provides a good guide (see also Figs 2.68 and 2.69):

Rigel (β Ori – bottom right-hand star) – about magnitude 0

Betelgeuse (α Ori – top left-hand star) – about magnitude 1

Bellatrix (γ Ori – top right-hand star) – about magnitude 1.5

Mintaka (δ Ori – right-hand star of the belt) – about magnitude 2

M42 (central "star" of the sword) – about magnitude 4

The brightest stars have to be accommodated on the scale by going to negative magnitudes. Thus we have:

Sirius (α CMa) – about magnitude –1.5

Canopus (α Car) – about magnitude –0.7

and

Rigil Kentaurus (α Cen) – about magnitude –0.3

The scale is set so that stars of magnitude 6 are just visible to a person with good eyesight on a clear moonless night from an excellent observing site. This corresponds roughly to the third versions of the photographs of each area of the sky shown in Chapter 2 (that is, to those captioned as "The stars visible from a brilliant site to an acute observer"). From an average site in developed countries, stars down to magnitude 4 or 5 should be visible (the second version of the photographs in Chapter 2); while from an urban site, often only stars of magnitude 1 or brighter will be seen (the first version of the photographs in Chapter 2). With a telescope from a reasonable site, fainter stars may be seen – down to magnitude 13 or 14 with an 8 inch (0.2 m) telescope, for example.

On the diagrams accompanying the photographs in Chapter 2, stars are shown in five magnitude ranges, by the sizes of the dots representing them. These ranges, in order of increasing dot size, are: magnitude 4.5 or fainter, magnitudes 4.5 to 3.5, magnitudes 3.5 to 2.5, magnitudes 2.5 to 1.5, and magnitude 1.5 or brighter.

In Section 3.2 the magnitudes of extended sources such as nebulae, star clusters and galaxies, are given. These are the integrated magnitudes – that is, the magnitude obtained by adding up all the light coming from all parts of the object. However the reader should be wary of taking these as an accurate guide to the ability to see the object. Firstly, if two objects have the same magnitude, but one covers a larger area of the sky than the other, then the former will appear in the telescope to be dimmer than the latter. The largest dimensions of the objects are also listed in Section 3.2, to give some indication of this effect. Secondly, many of the objects are not uniform in brightness. It may therefore be quite easy to see the brighter central parts of a dim galaxy, in comparison with a fairly uniform planetary nebula of the same integrated magnitude. Thirdly, and especially for the gaseous nebulae, the values may not be all that accurate, since they are quite difficult to measure. However all the Messier and Caldwell objects should be visible in a 3 inch (75 mm) or larger telescope.

3.1.2 Positions

Star hopping (Section 1.4) means that once one constellation has been found, other constellations and the objects within them can be found without knowing their precise positions. This section can therefore safely be ignored unless you wish to use other star maps and catalogues, or have a telescope that uses position indicators (also called setting circles).

The positions of stars and other objects in the sky is given by a coordinate system that is similar to longitude and latitude on the Earth. The equivalent sky coordinates are called *Right Ascension* and *Declination*.

Declination (abbreviated as Dec, symbol δ) is the direct equivalent of latitude. It is the angle between the object and the celestial equator. (The celestial equator is just the projection of the Earth's equator out into space: it is the line that divides the sky into northern and southern hemispheres.) Declination is measured in degrees, and is positive in the northern hemisphere and negative in the southern hemisphere. An object that has the same declination as the observer's latitude will pass through the zenith for that observer.

Right Ascension (abbreviated as RA, symbol α) is the equivalent of longitude, but with a couple of minor complications. Longitude is measured from a fixed point on the Earth – the position of the transit telescope at the original Royal Greenwich Observatory in London. But since the Earth is rotating in space, a fixed point on the Earth cannot be used to measure the positions of objects in the sky. A fixed point in the sky is needed for this latter purpose. The point chosen is where the Sun in its annual path around the sky moves from the southern hemisphere to the northern hemisphere (that is, where it crosses the celestial equator). That point is called the *First Point of Aries*, and it lies in the constellation of Pisces (see Fig. 2.30). Right ascension is measured eastward from the First Point of Aries. This is slightly different from longitude, which is measured to the east or to the west of the Greenwich meridian. The main complication, however, is that right ascension is measured in units of hours, minutes and seconds, not in degrees. One hour of right ascension is equal to $15°$, one minute to $15'$ and one second to $1''$. The reason for this is that the Earth rotates through $15°$ in one hour (or $15'$ in one minute, or $15''$ in one second).* It seems a trifle peculiar to start with, but since all star maps, catalogues etc. use the system, you quickly become accustomed to it.

It probably also seems peculiar to most readers that the First Point of Aries should be found in Pisces. The reason for this anomaly is that the Earth's spin has a wobble to it, called *precession*. This causes the Earth's rotational axis slowly to change its position in space. One effect of this is that the celestial poles (North and South) move with respect to the stars. The presence of Polaris (α UMi) close to the North Pole is thus a matter of chance, and the pole will move far away from Polaris in the next few thousand years. The second effect is that the celestial equator also moves with respect to the stars, and the position at which the Sun crosses it drifts around the sky. The First Point of Aries is thus not truly a fixed point like the Greenwich meridian. When the first star catalogues were devised a century or two BC, the First Point of Aries was indeed the start of the constellation of Aries, but it has now moved over $30°$ to be in Pisces. The movements of the celestial equator and the First Point of Aries mean that stars' positions measured with respect to these quantities change with time. For visual work, that change can be ignored. However, if you use a telescope then you may need to take it into account. Star maps and catalogues always give the positions for a specific time, known as the *epoch*. These are commonly chosen at 25-year intervals, with the next one being the year 2000. The positions recorded here for objects in Section 3.2 are for that epoch, and will be usable until at least 2010 or 2015. The method for correction of the precessional changes is beyond the scope of this book, but may be found in sources listed in the Bibliography (Appendix 3).

---

* Strictly, because of the Earth's movement around its orbit, one complete rotation through $360°$ takes 23 hours 56 minutes 6 seconds.

3.1.3 Variable Stars

The majority of stars are constant in their brightnesses. Some stars however vary in brightness; these are called variable stars, and they are often of great interest. Where variable stars reach sixth magnitude or brighter, or are of particular significance, they are listed for each constellation in Section 3.2. There are many types of variable star, and only the main ones are mentioned below; further details may be found in sources listed in the Bibliography (Appendix 3).

Stars may vary in brightness because of some intrinsic change, or because of external factors. The latter are called extrinsic variables. They are mostly eclipsing binaries. These are pairs of stars in orbit around each other, like a planet around the Sun, and whose orbital plane lies close to the line of sight from Earth. Although there are two stars, they are too close together to be seen separately even in the largest telescopes. When the orbital motion of one star causes it to pass in front of the other as we see it from Earth, it obscures (eclipses) all or part of the more distant star, and causes the total observed brightness to decrease. The main types of eclipsing binary star are Algol type, β Lyrae type and W Ursae Majoris type, with the differences arising from the relative separations of the stars involved.

Intrinsic variables include:

1. Cepheids, which have regular variations by up to one or two magnitudes over periods ranging from a few days to some tens of days. The stars are expanding and contracting in size. The archetype is δ Cep, and other related variables are the W Virginis stars and the RR Lyrae stars.

2. Long-period variables (Mira stars), which can change by many magnitudes, but on rather variable time scales of several years. The archetype is Mira (o Ceti).

3. R CrB stars, which show sudden decreases in brightness at irregular intervals, possibly owing to the formation of carbon dust (soot) in their outer atmospheres.

4. T Tau stars, which are irregular variables that are thought to be very young stars. The Herbig Ae and Be stars are probably similar.

5. Novae and supernovae, which are stars that brighten rapidly and by many magnitudes. The changes result from explosions, which, in the case of supernovae, completely disrupt the star. With a few exceptions (recurrent novae), they are not predictable, and are usually found in the early stages of their eruptions by

enthusiastic amateur astronomers who spend many hours searching the sky for these stars (and comets).

3.1.4 Star Clusters

Star clusters are groups of stars held together by their own gravitational attraction. The number of stars involved can range from a few to millions. The two main types are *open* or *galactic clusters* and *globular clusters*. Open clusters, such as the Pleiades (M 45 – Fig. 2.69) generally have less than a thousand members and are within our own galaxy. They are groups of relatively young stars recently formed from interstellar gas clouds. In a few hundred million years, such clusters will be disrupted by tides and no longer be identifiable.

Globular clusters, such as *ω* Cen (C 80 – Fig. 2.76) contain from tens of thousands to millions of stars. The stars are in a spherical (globular) distribution, and the clusters are in effect small satellite galaxies orbiting around our own Milky Way galaxy. Some at least of the globular clusters may have formed before the main galaxy, and they will remain as separate recognisable entities for many more thousands of millions of years.

Double stars and binary stars should also be mentioned here. Double stars are two stars seen to be close together in the sky, but in fact at very different distances from the Earth. Binary stars are two stars physically connected by gravity and in orbit around each other. Examples of double and binary stars are listed in Section 3.2 where these are detectable with the naked eye or through binoculars or small telescope.

3.1.5 Gaseous Nebulae

Gaseous nebulae are huge glowing clouds of gas in interstellar space. Although they look spectacular, especially on long-exposure images obtained using large telescopes, they are in fact of such low density that they would be counted as a vacuum here on Earth. They are generally associated with the beginnings and endings of stars' lives.

H II regions (pronounced aitch-two regions) are gas clouds such as the Orion nebula (M42). They can range up to hundreds of light years in size and contain tens of thousands times the amount of material in the Sun. They are regions of current star formation, and they glow because of hot young stars embedded within them which heat the gas to around 10,000°C. At an earlier stage, before they are heated by the newly formed stars, they are cold (–200°C), and often contain large quantities of dust. They then form dark absorbing regions, which can sometimes be seen when silhouetted against a brighter background (for example, the Coalsack in Crux).

Planetary nebulae (such as M57, the Ring nebula in Lyra), despite their name, have nothing at all to do with planets, except that a few of them look a bit planet-like in small telescopes. They are actually shells of material flung off at speeds of a few tens of kilometres per second by stars at their centres. The central stars also heat the nebulae until they glow sufficiently brightly to be seen. They contain much less material than the H II regions – usually only a small fraction of a solar mass. The central stars are at the ends of their lives and in a few hundred thousand years will contract down to become white dwarfs. The nebulae will expand and disperse into the interstellar medium. This is the probable fate for our Sun in another five to six thousand million years.

Supernova remnants (SNRs – for example M1, the Crab Nebula) can have a superficial similarity to planetary nebulae in some cases, but more often look like roughly circular collections of wispy filaments. They also result from stars at the ends of their lives, but stars much more massive than those that produce planetary nebulae. Large stars undergo spectacular explosions called supernovae which completely destroy the star. Most of the star's material is flung out into space by the explosion and forms the SNR. The energies involved are enormous. The Crab nebulae was observed to originate in a supernova nearly a thousand years ago, and yet its outer regions are still seen to be expanding at 10,000 km s^{-1} or more.

3.1.6 Galaxies

Galaxies are self-gravitating collections of stars, with the number of stars involved ranging from tens of millions to millions of millions. The solar system is towards the edge of a largish spiral galaxy which we see as the Milky Way (see Section 1.9). Most galaxies are too faint to be seen with the naked eye, though quite a number can be seen using binoculars or a small telescope. From their appearances on photographs, they are divided into Spiral, Elliptical and Irregular forms (S, E, and I in Section 3.2). Used visually however, small telescopes will usually only reveal the central brighter regions of the galaxies, whatever their nature, so that all the types in fact appear rather similar to each other. Examples include: M31 (the Andromeda galaxy, a large spiral galaxy viewed nearly edge-on, and just visible to the unaided eye), M87 (a very large elliptical galaxy in Virgo) and M82 (an irregular galaxy in Ursa Major).

3.2 Constellations

Andromeda (And, Andromedae)

Meaning and origin: female name. From Greek mythology – Andromeda was the daughter of Cassiopeia and Cepheus. She was rescued by Perseus after she had been chained to a rock to be eaten by Cetus (the latter is now usually interpreted as a whale, but a sea monster is more appropriate in this context). The constellation was first recorded in Ptolemy's *Almagest* in about AD 145.

Located on figures: 2.22, 2.24, 2.30, 2.37, 2.62, 2.111
Area: 720 square degrees
Number of stars visible to the naked eye under good conditions – 55

Named stars
Alpheratz or Sirrah – α And (co-equal brightest star with β And – magnitude 2.06)
Mirach – β And (co-equal brightest star with α And – magnitude 2.06)
Almach or Alamak – γ And (blue and gold visual double star, separation 10″)

Variable stars
R And (long-period variable or Mira-type – period 1.1 years – magnitude range 5.9 to 14.9)

Antlia (Ant, Antliae) – originally Antlia Pneumatica

Meaning and origin: Pump or Air Pump. Introduced as a new constellation by Lacaille in his star catalogue, *Coelum Australe Stelliferum* of 1763.

Located on figures: 2.117, 2.122
Area: 240 square degrees
Number of stars visible to the naked eye under good conditions – 9
Brightest star: α Ant, magnitude 4.25

Apus (Aps, Apodis)

Meaning and origin: Bee. Often alternatively interpreted as a Bird of Paradise. The confusion probably dates from Bode's *Uranographia* of 1801 wherein the constellation was named both as Apus (Bee) and as Avis Indica (Bird of the Indies). Introduced as a new constellation by Keyser and de Houtman around 1600.

Located on figures: 2.76, 2.82, 2.88
Area: 205 square degrees
Number of stars visible to the naked eye under good conditions – 9
Brightest star: α Aps, magnitude 3.83

Andromeda – Nebulae, galaxies and other objects of interest

Object	Name	NGC	Type	RA2000 H	m	Dec2000 degrees	Visual mag	Size (')
—	—	205*	Galaxy (E)	00	40	41.7	8.0	17
Messier 32	—	221*	Galaxy (E)	00	43	40.9	8.2	8
Messier 31	Andromeda	224	Galaxy (S)	00	43	41.3	3.5	180
Caldwell 28	—	752	Galactic Cluster	01	58	37.7	5.7	50
Caldwell 23	—	891	Galaxy (S)	02	23	42.4	10.0	14
Caldwell 22	Blue Snowball	7662	Planetary Nebula	23	26	42.6	9.2	2

* Companion to M31.

Apus – Nebulae, galaxies and other objects of interest

Object	Name	NGC	Type	RA2000 H	m	Dec2000 degrees	Visual mag	Size (')
Caldwell 107	—	6101	Globular Cluster	16	26	−72.2	9.3	11

Aquarius (Aqr, Aquarii) – a zodiacal constellation

Meaning and origin: Water Carrier. Possibly derives from Zeus pouring rain on to the Earth, or commemorates Ganymede, cup-bearer to the gods. The constellation was recorded in Ptolemy's *Almagest* in about AD 145, but probably pre-dates that – possibly back to Babylonian times (2000 BC).

Located on figures: <u>2.30</u>, 2.88, 2.94, 2.111
Area: 980 square degrees
Number of stars visible to the naked eye under good conditions – 56

Named stars
Sadalmelik – α Aqr (magnitude 2.96)
Sadalsuud – β Aqr (brightest star – magnitude 2.91)
Sadachbia – γ Aqr
Skat – δ Aqr
Albali – ε Aqr
Ancha – θ Aqr

Double stars
ζ Aqr (visual binary, magnitudes 4.3 and 4.5, separation 2″, period 850 years)

Aquila (Aql, Aquilae)

Meaning and origin: Eagle. From Greek mythology – Zeus changed into a black eagle in order to carry off Ganymede to become the cup-bearer to the gods. The constellation was recorded in Ptolemy's *Almagest* in about AD 145, but probably pre-dates that – possibly back to Babylonian times (2000 BC).

Located on figures: 2.37, <u>2.43</u>, <u>2.94</u>
Area: 650 square degrees

Number of stars visible to the naked eye under good conditions – 47

Named stars
Altair – α Aql (brightest star – magnitude 0.77; a part of the Summer Triangle)
Alshain – β Aql
Tarazed – γ Aql

Variable stars
η Aql (Cepheid variable – period 7.2 days – magnitude range 4.08 to 5.25)

Ara (Ara, Arae)

Meaning and origin: Altar. From Greek mythology – the altar used by the Olympian gods to swear allegiance to each other before their battle with the Titans. The constellation was first recorded in Ptolemy's *Almagest* in about AD 145.

Located on figures: <u>2.82</u>
Area: 235 square degrees
Number of stars visible to the naked eye under good conditions – 18
Brightest star: β Ara, magnitude 2.85

Aries (Ari, Arietis) – a zodiacal constellation

Meaning and origin: Ram. From Greek mythology – Phrixus and Helle flew off on a ram to escape their stepmother, Ino. Helle fell off and was drowned in the straits between Europe and Asia, giving her name to the Hellespont (now the Dardanelles). The ram bore Phrixus to safety, where with remarkable ingratitude he then sacrificed it. The ram's fleece was the golden fleece sought by Jason. The constellation was recorded in Ptolemy's

Aquarius – Nebulae, galaxies and other objects of interest								
Object	Name	NGC	Type	RA₂₀₀₀ H	m	Dec₂₀₀₀ degrees	Visual mag	Size (′)
Messier 72	—	6981	Globular Cluster	20	54	−12.5	9.4	6
Messier 73	—	6994	Open Cluster*	20	59	−12.6	8.9	3
Caldwell 55	Saturn	7009	Planetary Nebula	21	04	−11.4	8.3	2
Messier 2	—	7089	Globular Cluster	21	34	−0.8	6.5	13
Caldwell 63	Helix	7293	Planetary Nebula	22	30	−20.8	7.4	13

* Only four stars involved.

Ara – Nebulae, galaxies and other objects of interest

Object	Name	NGC	Type	RA$_{2000}$ H	m	Dec$_{2000}$ degrees	Visual mag	Size (')
Caldwell 82	—	6193	Open Cluster	16	41	−48.8	5.2	15
—	—	6250	Open Cluster	16	58	−45.8	5.9	8
—	—	IC 4651	Open Cluster	17	25	−50.0	6.9	12
Caldwell 81	—	6352	Globular Cluster	17	26	−48.4	8.2	7
Caldwell 86	—	6397	Globular Cluster	17	41	−53.7	5.7	26

Almagest in about AD 145, but probably pre-dates that – possibly back to Babylonian times (2000 BC).

Located on figures: 2.24, 2.30, 2.62, 2.111
Area: 440 square degrees
Number of stars visible to the naked eye under good conditions – 28

Named stars
Hamal – α Ari (brightest star – magnitude 2.00)
Sheratan – β Ari
Mesartim – γ Ari (visual binary star, separation 8.2″)

Double stars
γ Ari (magnitudes 4.7 and 4.7, separation 8″)
ε Ari (visual binary, magnitudes 5.2 and 5.5, separation 1.5″)

Auriga (Aur, Aurigae)

Meaning and origin: Charioteer. From Greek mythology – a son of Hephaestos (Vulcan) and a cripple from birth, Erichthonius (Auriga) was reputed to have been the inventor of the chariot. The constellation was recorded in Ptolemy's *Almagest* in about AD 145, but probably pre-dates that – possibly back to Babylonian times (2000 BC).

Located on figures: 2.22, 2.24, 2.62, 2.69
Area: 655 square degrees
Number of stars visible to the naked eye under good conditions – 50

Named stars
Capella – α Aur (brightest star – magnitude 0.08)
Menkalinan – β Aur (variable – see below)

Variable stars
β Aur (Algol type eclipsing binary star – period 4 days – magnitude range 1.9 to 2.0)
ε Aur (eclipsing binary – period 27 years – magnitude range 3.7 to 4.5)
ζ Aur (eclipsing binary – period 2.7 years – magnitude range 5.0 to 5.6)
ψ^1 Aur (irregular variable – magnitude range 4.5 to 5.6)

Boötes (Boö, Boötis)

Meaning and origin: Herdsman. Represented on some old star maps as holding the hunting dogs (Canes Venatici) in leash and driving the great bear (Ursa Major) around the sky. The constellation was recorded in Ptolemy's *Almagest* in about AD 145, but probably pre-dates that – possibly back to Babylonian times (2000 BC).

Located on figures: 2.13, 2.43, 2.48
Area: 905 square degrees
Number of stars visible to the naked eye under good conditions – 52

Named stars
Arcturus – α Boö (The brightest star – magnitude 0.06. The fourth brightest star in the night sky and the brightest star in the northern hemisphere.)

Auriga – Nebulae, galaxies and other objects of interest

Object	Name	NGC	Type	RA$_{2000}$ H	m	Dec$_{2000}$ degrees	Visual mag	Size (')
Caldwell 31	Flaming Star	IC 405	Emission Nebula	05	16	34.3	≈ 7	3
—	—	1893	Open Cluster	05	23	33.4	7.5	11
Messier 38	—	1912	Open Cluster	05	29	35.8	6.4	21
Messier 36	—	1960	Open Cluster	05	36	34.1	6.0	12
Messier 37	—	2099	Open Cluster	05	52	32.6	5.6	24

Nekkar – β Boö

Seginus – γ Boö

<u>Izar</u>, Mizar or Pulcherrima – ε Boö (visual binary star, separation 2.9″; Mizar is also the name of ζ UMa)

Muphrid – η Boö

Alkalurops – μ Boö

Variable stars

W Boö (semi-regular variable, magnitude range 4.5 to 5.5, time scale a few months)

Double stars

ζ Boö (visual binary, magnitudes 4.4 and 4.6, separation 1″, period 120 years)

Cælum (Cae, Cæli) – originally Cæla Sculptoris

Meaning and origin: Sculptor's chisel or engraving tool. Introduced as a new constellation by Lacaille in his star catalogue, *Coelum Australe Stelliferum* of 1763.

Located on figures: <u>2.100</u>

Area: 125 square degrees

Number of stars visible to the naked eye under good conditions – 4. This is the sparsest of the constellations

Brightest star: α Cae, magnitude 4.45

Camelopardalis (Cam, Camelopardalis), also called Camelopardus

Meaning and origin: Giraffe. Introduced as a new constellation by Plancius on his star globe of 1613.

Located on figures: <u>2.22</u>, <u>2.24</u>, <u>2.62</u>

Area: 755 square degrees

Number of stars visible to the naked eye under good conditions – 43

Brightest star: β Cam, magnitude 4.08

Variable stars

VZ Cam (semi-regular variable, magnitude range 4.5 to 5.3, time scale 20 days)

Cancer (Cnc, Cancri) – a zodiacal constellation

Meaning and origin: Crab. From Greek mythology – the crab that attacked Heracles (Hercules) and was killed during the latter's fight with the sea serpent (Hydra). The constellation was recorded in Ptolemy's *Almagest* in about AD 145, but probably pre-dates that – possibly back to Babylonian times (2000 BC).

Located on figures; <u>2.55</u>, <u>2.69</u>, 2.122

Area: 505 square degrees

Number of stars visible to the naked eye under good conditions – 23

Brightest star: β Cnc, magnitude 3.52

Named stars

Acubens – α Cnc

Asellus Borealis – γ Cnc

Asellus Australis – δ Cnc

Tegmeni – ζ Cnc (A multiple visual binary system. The brightest component is of magnitude 5.6, and it has a magnitude 6.0 companion about 1″ away from it in a 60-year long orbit. The third component is of magnitude 6.2 and is itself a binary with a separation of about

Boötes – Nebulae, galaxies and other objects of interest								
Object	Name	NGC	Type	RA$_{2000}$ H	m	Dec$_{2000}$ degrees	Visual mag	Size (′)
Caldwell 45	—	5248	Galaxy (S)	13	38	8.9	10.2	7

Camelopardalis – Nebulae, galaxies and other objects of interest								
Object	Name	NGC	Type	RA$_{2000}$ H	m	Dec$_{2000}$ degrees	Visual mag	Size (′)
Caldwell 5	—	IC 342	Galaxy (S)	03	47	68.1	9.1	18
—	—	1502	Open Cluster	04	08	62.3	5.7	8
Caldwell 7	—	2403	Galaxy (S)	07	37	65.6	8.4	18

Cancer – Nebulae, galaxies and other objects of interest				RA$_{2000}$		Dec$_{2000}$	Visual	Size
Object	Name	NGC	Type	H	m	degrees	mag	(')
Messier 44	Praesepe	2632	Open Cluster	08	40	20.0	3.1	95
Messier 67	—	2682	Open Cluster	08	50	11.8	6.9	30
Caldwell 48	—	2775	Galaxy (S)	09	10	7.0	10.3	5

0.25″. This latter system orbits the former at a distance of about 6″ with a period in excess of 1000 years.)

Canes Venatici (CVn, Canum Venaticorum)

Meaning and origin: Hunting Dogs (named Chara and Asterion). Introduced as a new constellation by Hevelius in his star atlas *Firmamentum Sobiescianum* of 1690.

Located on figures: 2.13, 2.48
Area: 465 square degrees
Number of stars visible to the naked eye under good conditions – 14

Named stars
Cor Caroli – α CVn (The brightest star, magnitude 2.89. This is the prototype of the α CVn-type of variable star whose variations arise from surface features combined with rotation. It is also a visual double star with a fifth magnitude companion about 20″ away. The name means Charles' Heart, and probably recognises the foundation of the Royal Greenwich Observatory by Charles II, alternatively however it may have arisen from the beheading of Charles I.)
Chara – β CVn

Canis Major (CMa, Canis Majoris)

Meaning and origin: Large Dog. One of Orion's two dogs (see also Ursa Minor). The constellation was first recorded in Ptolemy's *Almagest* in about AD 145.

Located on figures: 2.69, 2.100, 2.117
Area: 380 square degrees
Number of stars visible to the naked eye under good conditions – 54

Named stars
Sirius – α CMa (The brightest star and the brightest star in the night sky, magnitude –1.47. It is also a visual binary with an eighth-magnitude white dwarf as a companion. The separation ranges from 3″ to 11″, with an orbital period of 50 years.)
Mirzam – β CMa
Wezen – δ CMa
Adhara – ε CMa
Furud – ζ CMa
Aludra – η CMa

Variable stars
ω CMa (emission line star, magnitude range 3.5 to 4.3)
EW CMa (emission line star, magnitude range 4.4 to 4.8)

Canes Venatici – Nebulae, galaxies and other objects of interest				RA$_{2000}$		Dec$_{2000}$	Visual	Size
Object	Name	NGC	Type	H	m	degrees	mag	(')
—	—	4214	Galaxy (I)	12	16	36.3	9.7	8
Caldwell 26	—	4244	Galaxy (S)	12	18	37.8	10.2	16
Messier 106	—	4258	Galaxy (S)	12	19	47.3	8.3	18
Caldwell 21	—	4449	Galaxy (I)	12	28	44.1	9.4	5
—	—	4490	Galaxy (S)	12	31	41.6	9.8	6
Caldwell 32	—	4631	Galaxy (S)	12	42	32.5	9.3	15
Messier 94	—	4736	Galaxy (S)	12	51	41.1	8.2	11
Caldwell 29	—	5005	Galaxy (S)	13	11	37.1	9.8	5
Messier 63	Sunflower	5055	Galaxy (S)	13	16	42.0	8.6	12
Messier 51	Whirlpool	5194	Galaxy (S)	13	30	47.2	8.4	11
Messier 3	—	5272	Globular Cluster	13	42	28.4	6.4	16

Canis Major – Nebulae, galaxies and other objects of interest								
Object	Name	NGC	Type	RA2000		Dec2000	Visual	Size
				H	m	degrees	mag	(')
Messier 41	—	2287	Open Cluster	06	47	−20.7	4.5	38
Caldwell 58	—	2360	Open Cluster	07	18	−15.6	7.2	13
Caldwell 64	—	2362	Open Cluster	07	19	−25.0	4.1	8

UW CMa (eclipsing binary, magnitude range 4.8 to 5.3, period 4.4 days)

Canis Minor (CMi, Canis Minoris)

Meaning and origin: Small Dog. One of Orion's two dogs (see also Ursa Major). The constellation was first recorded in Ptolemy's *Almagest* in about AD 145.

Located on figures: 2.55, 2.69, 2.117
Area: 185 square degrees
Number of stars visible to the naked eye under good conditions – 13

Named stars
Procyon – α CMi (The brightest star, magnitude 0.34. It is also a visual binary with an eleventh-magnitude white dwarf as a companion. The separation ranges from 2″ to 4″, with an orbital period of about 41 years.)
Gomeisa – β CMi

Capricornus (Cap, Capricorni) – the most inconspicuous of the zodiacal constellations

Meaning and origin: Goat. Possibly representing Pan from Greek mythology. The constellation was recorded in Ptolemy's *Almagest* in about AD 145, but probably predates that – possibly back to Babylonian times (2000 BC).

Located on figures: 2.43, 2.88, 2.94, 2.111
Area: 415 square degrees
Number of stars visible to the naked eye under good conditions – 32

Named stars
Algedi or Giedi – α Cap (a naked-eye double star, separation 6.2′, magnitudes 3.56 and 4.24; both components are also visual doubles with faint companions visible through a telescope)
Dabih – β Cap
Nashira – γ Cap
Deneb Algedi – δ Cap (the brightest star, magnitude 2.83)

Carina (Car, Carinae)

Meaning and origin: Keel. Originally this constellation, along with Puppis (poop or stern) and Vela (sails), formed the constellation Argo (ship – also called Argo Navis and one of Ptolemy's constellations). Argo, in Greek mythology, was Jason's ship as used by the argonauts during the hunt for the golden fleece. In Lacaille's star catalogue *Coelum Australe Stelliferum* of 1763, Argo was subdivided into the three smaller constellations for convenience.

Located on figures: 2.76, 2.100, 2.117
Area: 495 square degrees.
Number of stars visible to the naked eye under good conditions – 75

Named stars
Canopus – α Car (the brightest star and the second brightest star in the night sky, magnitude –0.73)
Miaplacidus – β Car
Avior – ε Car
Aspidiske – ι Car

Variable stars
η Car (irregular variable, magnitude range 5.8 to 8, but reached magnitude –0.8 in 1843)

Capricornus – Nebulae, galaxies and other objects of interest								
Object	Name	NGC	Type	RA2000		Dec2000	Visual	Size
				H	m	degrees	mag	(')
Messier 30	—	7099	Globular Cluster	21	40	−23.2	7.5	11

R Car (long-period variable, magnitude range 4 to 11, period about 10 months)

S Car (long period variable, magnitude range 4.4 to 10, period about 5 months)

I Car (Cepheid, magnitude range 3.2 to 4.2, period 36 days)

ZZ Car (Cepheid, magnitude range 3.3 to 4.3, period 36 days)

Cassiopeia (Cas, Cassiopeiae)

Meaning and origin: female name. From Greek mythology – Cassiopeia was an Ethiopian queen, wife of Cepheus and mother of Andromeda. As punishment for boasting of her own and of her daughter's beauty, a monster (Cetus) was sent to plague the country. To save themselves, Andromeda had to be sacrificed to the monster, however she was rescued by Perseus who turned the monster to stone by revealing the head of Medusa. The constellation was first recorded in Ptolemy's *Almagest* in about AD 145.

Located on figures: 2.22, 2.24, 2.62
Area: 600 square degrees

Numbers of stars visible to the naked eye under good conditions – 52

Named stars
Schedar – α Cas (The brightest star – magnitude 2.24. Variable – see below.)
Caph – β Cas (magnitude 2.25)
Cih – γ Cas (Magnitude 2.65. Variable – see below. The brightest of the Be stars; hot stars with prominent emission lines in their spectra. A visual double star with an eleventh magnitude companion, separation 2.1″.)
Ruchbah – δ Cas

Variable stars
α Cas (irregular variable, magnitude range 2.2 to 2.8)
γ Cas (irregular variable, magnitude range 1.6 to 3.1)
ρ Cas (semi-regular variable, magnitude range 4 to 6)
R Cas (long-period variable, magnitude range 4.5 to 13, period 14 months)
YZ Cas (eclipsing binary, magnitude range 5.6 to 6.1, period 4.5 days)
SU Cas (Cepheid, magnitude range 5.6 to 6.1, period 2 days)

Carina – Nebulae, galaxies and other objects of interest

Object	Name	NGC	Type	RA₂₀₀₀ H	m	Dec₂₀₀₀ degrees	Visual mag	Size (′)
Caldwell 96	—	2516	Open Cluster	07	58	−60.9	3.8	30
—	—	2808	Globular Cluster	09	12	−64.9	6.3	14
Caldwell 90	—	2867	Planetary Nebula	09	21	−58.3	9.7	0.2
—	—	3114	Open Cluster	10	03	−60.1	4.2	35
Caldwell 102	Southern Pleiades	IC 2602	Open Cluster	10	43	−64.4	1.9	50
Caldwell 92	Eta Carina	3372	Emission Nebula	10	44	−59.9	2.5	120
Caldwell 91	—	3532	Open Cluster	11	06	−58.7	3.0	55

Cassiopeia – Nebulae, galaxies and other objects of interest

Object	Name	NGC	Type	RA₂₀₀₀ H	m	Dec₂₀₀₀ degrees	Visual mag	Size (′)
Caldwell 17	—	147	Galaxy (E)	00	33	48.5	9.3	13
Caldwell 18	—	85	Galaxy (E)	00	39	48.3	9.2	12
—	—	225	Open Cluster	00	43	61.8	7.0	12
Caldwell 13	—	457	Open Cluster	01	19	58.3	6.4	13
Caldwell 8	—	559	Open Cluster	01	30	63.3	9.5	5
Messier 103	—	581	Open Cluster	01	33	60.7	7.4	6
Caldwell 10	—	663	Open Cluster	01	46	61.3	7.1	16
Caldwell 11	Bubble	7635	Emission Nebula	23	21	61.2	8.5	15
Messier 52	—	7654	Open Cluster	23	24	61.6	6.9	13
—	—	7789	Open Cluster	23	57	56.7	6.7	16

Centaurus (Cen, Centauri)

Meaning and origin: Centaur. From Greek mythology – a beast consisting of a human head and torso set upon a horse's body. Supposedly originally fathered by Centaurus (son of Ixion and Nephele) upon mares belonging to Magnes. In fact the beast is derived from the first encounters of the chariot-riding Greeks with horses being ridden by invaders from the north. The constellation was recorded in Ptolemy's *Almagest* in about AD 145, but probably pre-dates that – possibly back to Babylonian times (2000 BC).

Located on figures: <u>2.76</u>, 2.82, 2.100, 2.117
Area: 1060 square degrees.
Number of stars visible to the naked eye under good conditions – 101. This is the most populated of the constellations

Named stars
Rigil Kentaurus – α Cen (The brightest star and the third brightest star in the night sky, magnitude –0.27. A visual double with a separation ranging from 2″ to 25″ and an orbital period of 80 years. The components have magnitudes of 0.0 and 1.7. A tenth-magnitude third member of the system, located about 1.5° south of α Cen, is the closest star to the Earth after the Sun. This star is called Proxima Centauri, and its distance from us is 1.31 pc (4.26 ly).)
Agena – β Cen (visual double star, magnitudes 0.7 and 3.8, separation 1.5″)
Menkent – θ Cen

Variable stars
μ Cen (emission line star, magnitude range 2.8 to 3.6)
o^1 Cen (semi-regular variable, magnitude range 4.7 to 5.5, time scale 6 months)

Double stars
α Cen (see above)
β Cen (see above)
γ Cen (visual binary, magnitudes 3.0 and 3.0, separation 1″, period 80 years)

Cepheus (Cep, Cephei)

Meaning and origin: male name. From Greek mythology – an Ethiopian king, husband of Cassiopeia and father of Andromeda (see above). The constellation was first recorded in Ptolemy's *Almagest* in about AD 145.

Located on figures: <u>2.22</u>, 2.37
Area: 590 square degrees
Number of stars visible to the naked eye under good conditions – 57

Named stars
Alderamin – α Cep (the brightest star, magnitude 2.41)
Alfirk – β Cep (variable – see below)
<u>Alrai</u> or Errai – γ Cep
Garnet – μ Cep (see below)
Kurhah – ζ Cep

Variable stars
β Cep (Archetype of the β Cep-type variable stars. These are also known as the β CMa variables. They have a small magnitude range and periods of a few hours.)
δ Cep (Archetype of the classical Cepheid variable stars. Magnitude range 3.6 to 4.3, period 5.4 days.)
μ Cep (Semi-regular variable star. Magnitude range 3.6 to 5.1 on a time scale of a year or so. The reddest naked eye star in the northern sky, and so also known as the Garnet Star.)
VV Cep (eclipsing binary, magnitude range 4.8 to 5.5, period 20 years)

Centaurus – Nebulae, galaxies and other objects of interest

Object	Name	NGC	Type	RA2000 H	m	Dec2000 degrees	Visual mag	Size (′)
Caldwell 97	—	3766	Open Cluster	11	36	–61.6	5.3	12
Caldwell 100	—	IC 2944	Open Cluster	11	37	–63.0	4.5	15
—	Blue Planetary	3918	Planetary Nebula	11	50	–57.2	8.4	12
Caldwell 83	—	4945	Galaxy (S)	13	05	–49.5	8.6	20
Caldwell 77	Cen A	5128	Galaxy (S)	13	26	–43.0	7.0	18
Caldwell 80	ω Centauri	5139	Globular Cluster	13	27	–47.5	3.7	36
Caldwell 84	—	5286	Globular Cluster	13	46	–51.4	7.6	9
—	—	5316	Open Cluster	13	54	–61.9	6.0	14
—	—	5460	Open Cluster	14	08	–48.3	5.6	25
—	—	5617	Open Cluster	14	30	–60.7	6.3	10
—	—	5662	Open Cluster	14	35	–56.6	5.5	12

Cepheus – Nebulae, galaxies and other objects of interest

Object	Name	NGC	Type	RA₂₀₀₀ H	m	Dec₂₀₀₀ degrees	Visual mag	Size (')
Caldwell 2	—	40	Planetary Nebula	00	13	72.5	10.7	0.6
Caldwell 1	—	188	Open Cluster	00	44	85.3	8.1	14
Caldwell 12	—	6946	Galaxy (S)	20	35	60.2	8.9	11
Caldwell 4	—	7023	Reflection Nebula	21	02	68.2	7.0	18
—	—	IC 1369	Open Cluster	21	39	57.5	3.5	50
Caldwell 9	Cave	Sh2-155	Emission Nebula	22	57	62.6	≈ 9	50

Cetus (Cet, Ceti)

Meaning and origin: Whale or Sea Monster. From Greek mythology – the monster sent in punishment for Cassiopeia's boasting, which was turned to stone by Perseus. The constellation was recorded in Ptolemy's *Almagest* in about AD 145, but probably pre-dates that – possibly back to Babylonian times (2000 BC).

Located on figures: 2.106, 2.111
Area: 1230 square degrees
Number of stars visible to the naked eye under good conditions – 55

Named stars
Menkab or Menkar – α Cet
Deneb Kaitos or Diphda – β Cet (the brightest star, magnitude 2.04)
Baten Kaitos – ζ Cet
Mira – o Cet (variable – see below)

Variable stars
o Cet (The archetypal long-period variable star, which are therefore also known as Mira variables. Magnitude range about 3.5 to 8.5 over a period of about one year.)

Chamaeleon (Cha, Chamaelontis)

Meaning and origin: Chamaeleon. Introduced as a new constellation by Keyser and de Houtman around 1600.

Located on figures: 2.76, 2.88, 2.100
Area: 130 square degrees
Number of stars visible to the naked eye under good conditions – 12
Brightest star – α Cha – magnitude 4.06

Circinus (Cir, Circini)

Meaning and origin: Geometer's compasses. Introduced as a new constellation by Lacaille in his star catalogue, *Coelum Australe Stelliferum* of 1763.

Located on figures: 2.76, 2.82
Area: 95 square degrees
Number of stars visible to the naked eye under good conditions – 9
Brightest star – α Cir – magnitude 3.17

Cetus – Nebulae, galaxies and other objects of interest

Object	Name	NGC	Type	RA₂₀₀₀ H	m	Dec₂₀₀₀ degrees	Visual mag	Size (')
Caldwell 62	—	247	Galaxy (S)	00	47	−20.8	8.9	20
Caldwell 56	—	246	Planetary Nebula	00	47	−11.9	8.0	4
Caldwell 51	—	IC 1613	Galaxy (I)	01	05	2.1	9.3	12
Messier 77	—	1068	Galaxy (S)	02	43	0.0	8.8	7

Chamaeleon – Nebulae, galaxies and other objects of interest

Object	Name	NGC	Type	RA₂₀₀₀ H	m	Dec₂₀₀₀ degrees	Visual mag	Size (')
Caldwell 109	—	3195	Planetary Nebula	10	10	−80.9	11.6	0.6

Circinus – Nebulae, galaxies and other objects of interest								
Object	Name	NGC	Type	RA2000 H	m	Dec2000 degrees	Visual mag	Size (')
Caldwell 88	—	5823	Open Cluster	15	06	−55.6	7.9	10

Columba (Col, Columbae) – originally Columba Noachii

Meaning and origin: Dove. The dove sent from Noah's Ark to find land. Introduced as a new constellation by Plancius in 1592.

Located on figures: 2.100, 2.117
Area: 270 square degrees
Number of stars visible to the naked eye under good conditions – 22

Named stars
Phact – α Col (the brightest star, magnitude 2.63)
Wazn – β Col

Coma Berenices (Com, Comae Berenices)

Meaning and origin: Berenice's Hair. Queen Berenice of Egypt pledged her long and beautiful hair for the safe return of her husband from a war with the Assyrians. The constellation probably dates back to 200 BC, but was only formally defined by Mercator in 1551.

Located on figures: 2.13, 2.48, 2.55
Area: 385 square degrees
Number of stars visible to the naked eye under good conditions – 23
Brightest star – β Com – magnitude 4.26

Corona Australis (CrA, Coronae Australis)

Meaning and origin: Southern Crown. The constellation was first recorded in Ptolemy's *Almagest* in about AD 145.

Located on figures: 2.82, 2.94
Area: 130 square degrees
Number of stars visible to the naked eye under good conditions – 20
Brightest stars – α CrA and β CrA – both of magnitude 4.10

Columba – Nebulae, galaxies and other objects of interest								
Object	Name	NGC	Type	RA2000 H	m	Dec2000 degrees	Visual mag	Size (')
Caldwell 73	—	1851	Globular Cluster	05	14	−40.1	7.3	11

Coma Berenices – Nebulae, galaxies and other objects of interest								
Object	Name	NGC	Type	RA2000 H	m	Dec2000 degrees	Visual mag	Size (')
Messier 98	—	4192	Galaxy (S)	12	14	14.9	10.1	10
Messier 99	—	4254	Galaxy (S)	12	19	14.4	9.8	5
Messier 100	—	4321	Galaxy (S)	12	23	15.8	9.4	7
Messier 85	—	4382	Galaxy (S)	12	25	18.2	9.2	7
Messier 88	—	4501	Galaxy (S)	12	32	14.4	9.5	7
Messier 91	—	4548	Galaxy (S)	12	35	14.5	10.2	5
Caldwell 38	—	4565	Galaxy (S)	12	36	26.0	9.6	16
Caldwell 36	—	4559	Galaxy (S)	12	36	28.0	9.9	11
—		4725	Galaxy (S)	12	50	25.5	9.2	11
Messier 64	Black-eye	4826	Galaxy (S)	12	57	21.7	8.5	9
Caldwell 35	—	4889	Galaxy (E)	13	00	28.0	11.4	3
Messier 53	—	5024	Globular Cluster	13	13	18.2	7.7	13

Corona Australis – Nebulae, galaxies and other objects of interest								
Object	Name	NGC	Type	RA$_{2000}$ H	m	Dec$_{2000}$ degrees	Visual mag	Size (')
Caldwell 78	—	6541	Globular Cluster	18	08	–43.7	6.6	13
Caldwell 68	R CrA	6729	Reflection Nebula	19	02	–37.0	≈ 11	1

Double stars
γ CrA (visual binary, magnitudes 4.7 and 5.0, separation 1.3″, period 125 years)

Corona Borealis (CrB, Coronae Borealis)

Meaning and origin: Northern Crown. From Greek mythology – this is the crown given to Ariadne by Dionysus (Bacchus) upon their marriage. The constellation was first recorded in Ptolemy's *Almagest* in about AD 145.

Located on figures: 2.13, 2.43, 2.48
Area: 180 square degrees
Number of stars visible to the naked eye under good conditions – 20

Named stars
Alphecca – α CrB (brightest star, magnitude 2.23)
Nusakan – β CrB

Variable stars
R CrB (Archetype of the R CrB variables. Magnitude range 6 to 14 on time scales of a few months.)
T CrB (Recurrent nova. Magnitude range 2 to 10, interval between outbursts 80 years).

Corvus (Crv, Corvi)

Meaning and origin: Crow. From Greek mythology – one version has it that Apollo left a crow (then a white bird) to guard Coronis while he was in Delphi. Coronis' infidelity was not prevented by the crow, who was turned black along with all his descendants as a punishment for the failure. Another suggestion is that it is the crow sent by Apollo to fetch water for a sacrifice. The crow delayed to eat figs, and was condemned to everlasting thirst by Apollo – hence the rasping call of the bird. The constellation was first recorded in Ptolemy's *Almagest* in about AD 145.

Located on figures: 2.76, 2.117, 2.122
Area: 185 square degrees
Number of stars visible to the naked eye under good conditions – 10

Named stars
Alchiba – α Crv
Gienah or Minkar – γ Crv (the brightest star, magnitude 2.6; also the name of ε Cyg)
Algorab – δ Crv

Crater (Crt, Crateris)

Meaning and origin: Cup. From Greek mythology – possibly the cup of Apollo, or the wine goblet of Dionysus (Bacchus). The constellation was first recorded in Ptolemy's *Almagest* in about AD 145.

Located on figures: 2.117, 2.122
Area: 280 square degrees
Number of stars visible to the naked eye under good conditions – 9
Brightest star – δ Crt – magnitude 3.56

Named stars
Alkes – α Crt

Crux (Cru, Crucis) – also known as Crux Australis

Meaning and origin: a cross. Recognised from early times, but regarded as a part of Centaurus until 1598 when it was

Corvus – Nebulae, galaxies and other objects of interest								
Object	Name	NGC	Type	RA$_{2000}$ H	m	Dec$_{2000}$ degrees	Visual mag	Size (')
Caldwell 60	Antennae	4038	Galaxy (S)	12	02	–18.9	10.7	3
Caldwell 61	Antennae	4039	Galaxy (S)	12	02	–18.9	10.7	3

shown as a separate constellation by Plancius on his star globe.

Located on figures: <u>2.76</u>, <u>2.117</u>
Area: 70 square degrees – the smallest of the constellations.
Number of stars visible to the naked eye under good conditions – 20

Named stars
Acrux – α Cru (the brightest star, magnitude 0.87; visual binary, magnitudes 1.3 and 1.7, separation 4.5″)
Gacrux – γ Cru

Cygnus (Cyg, Cygni) – also known as the Northern Cross

Meaning and origin: Swan. From Greek mythology – either the swan into which Zeus (Jupiter) metamorphosed himself in order to ravish Leda, or the boy Cygnus who was metamorphosed into a swan to search for the body of his friend Phaëthon. Phaëthon had been drowned in the river Eridanus by Zeus (Jupiter) after losing control of the chariot of the Sun. The constellation was first recorded in Ptolemy's *Almagest* in about AD 145.

Located on figures: 2.22, 2.30, <u>2.37</u>, 2.43
Area: 805 square degrees
Number of stars visible to the naked eye under good conditions – 80

Named stars
Deneb – α Cyg (the brightest star, magnitude 1.26; a part of the Summer Triangle)
Albireo – β Cyg (blue and gold double star, separation 35″, magnitudes 3.2 and 5.4)
Sadr – γ Cyg
Gienah – ε Cyg (also the name of γ Crv)

Variable stars
χ Cyg (long-period variable, magnitude range 3.3 to 14.2, period 1.1 years)
P Cyg (η Car-type variable, magnitude 4.88, time scale for changes is tens of years)

Delphinus (Del, Delphini)

Meaning and origin: Dolphin. The dolphin that reputedly saved the life of the musician Arion when he jumped off a ship to escape sailors plotting to kill him and steal prizes won in a competition. Alternatively the messenger of Poseidon (Neptune). The constellation was recorded in Ptolemy's *Almagest* in about AD 145, but probably pre-dates that – possibly back to Babylonian times (2000 BC).

Located on figures: <u>2.30</u>, <u>2.37</u>, 2.94
Area: 190 square degrees
Number of stars visible to the naked eye under good conditions – 10

Crux – Nebulae, galaxies and other objects of interest							
Object	Name	NGC	Type	RA₂₀₀₀ H m	Dec₂₀₀₀ degrees	Visual mag	Size (′)
Caldwell 98	—	4609	Open Cluster	12 42	−63.0	6.9	5
Caldwell 99	Coalsack	—	Absorption Nebula	12 53	−63.0	—	350
Caldwell 94	Jewel Box	4755	Open Cluster	12 54	−60.3	4.2	10

Cygnus – Nebulae, galaxies and other objects of interest							
Object	Name	NGC	Type	RA₂₀₀₀ H m	Dec₂₀₀₀ degrees	Visual mag	Size (′)
Caldwell 15	Blinking	6826	Planetary Nebula	19 45	50.5	9.8	2
Caldwell 27	Crescent	6888	Emission Nebula	20 12	38.4	≈11	20
Messier 29	—	6913	Open Cluster	20 24	38.5	6.6	7
Caldwell 34	Veil (W)	6960	Supernova Remnant	20 46	30.7	8.0	70
Caldwell 33	Veil (E)	6992 / 5	Supernova Remnant	20 57	31.5	8.0	60
Caldwell 20	North America	7000	Emission Nebula	20 59	44.3	5.0	120
Messier 39	—	7092	Open Cluster	21 32	48.4	4.6	32
Caldwell 19	Cocoon	IC 5146	Emission Nebula	21 54	47.3	7.2	12

Named stars
Sualocin – α Del (the brightest star, magnitude 3.77)
Rotanev – β Del (magnitude 3.78)

Double stars
γ Del (magnitudes 4.5 and 5.4, separation 10″)

Dorado (Dor, Doradus)

Meaning and origin: literally Golden-coloured fish – from the colour taken on by dying coryphenes (dolphins). Nowadays it is usually taken to be a Swordfish. Introduced as a new constellation by Keyser and de Houtman around 1600.

Located on figures: 2.88, 2.100
Area: 180 square degrees
Number of stars visible to the naked eye under good conditions – 15
Brightest star – α Dor – magnitude 3.26

Variable stars
β Dor (Cepheid, magnitude range 3.8 to 4.6, period 10 days)
R Dor (semi-regular variable, magnitude range 4.5 to 6.5, time scale 1 year)

Draco (Dra, Draconis)

Meaning and origin: Dragon. Many possible legends exist to explain this constellation. One is the Babylonian story of Bel and the Dragon, another the dragon Ladon which guarded the golden apples of the Hesperides and which was killed by Heracles (Hercules). The constellation was recorded in Ptolemy's *Almagest* in about AD 145, but probably pre-dates that – possibly back to Babylonian times (2000 BC).

Located on figures: (2.13 & 2.37), 2.22, 2.62
Area: 1085 square degrees
Number of stars visible to the naked eye under good conditions – 75

Named stars
Thuban – α Dra
Alwaid or Rastaban – β Dra
Eltamin or Etamin – γ Dra (the brightest star, magnitude 2.22)
Altais – δ Dra
Aldhibain – η Dra
Edasich – ι Dra
Giausar – λ Dra
Alrakis – μ Dra
Grumium – ζ Dra

Delphinus – Nebulae, galaxies and other objects of interest							
Object	Name	NGC	Type	RA$_{2000}$ H m	Dec$_{2000}$ degrees	Visual mag	Size (')
Caldwell 47	—	6934	Globular Cluster	20 34	7.4	8.9	6
Caldwell 42	—	7006	Globular Cluster	21 02	16.2	10.6	3

Dorado – Nebulae, galaxies and other objects of interest							
Object	Name	NGC	Type	RA$_{2000}$ H m	Dec$_{2000}$ degrees	Visual mag	Size (')
Large Magellanic Cloud	—	—	Galaxy (S?)	05 20	−68.0	0.1	600
Caldwell 103	Tarantula	2070	Emission Nebula	05 39	−69.1	8.2	40

Draco – Nebulae, galaxies and other objects of interest							
Object	Name	NGC	Type	RA$_{2000}$ H m	Dec$_{2000}$ degrees	Visual mag	Size (')
Caldwell 3	—	4236	Galaxy (S)	12 17	69.5	9.7	19
Messier 102	—	5866	Galaxy (S)	15 07	55.8	10.0	5
Caldwell 6	Cat's Eye	6543	Planetary Nebula	17 59	66.6	8.8	6

Double stars
ν Dra (magnitudes 4.9 and 4.9, separation 62″)
40/41 Dra (magnitudes 5.6 and 6.0, separation 19″)

Equuleus (Equ, Equulei)

Meaning and origin: Little Horse. From Greek mythology – either it represents Thea who was turned into a mare while pregnant to avoid the wrath of her father, or the horse given to Castor by Hermes (Mercury). The constellation was first recorded in Ptolemy's *Almagest* in about AD 145.

Located on figures: 2.30, 2.37, 2.94
Area: 70 square degrees
Number of stars visible to the naked eye under good
   conditions – 5

Named stars
Kitalpha – α Equ (the brightest star, magnitude 3.92)

Eridanus (Eri, Eridani)

Meaning and origin: River. From Greek mythology – the river in which Phaëthon drowned after losing control of the chariot of the Sun (see also Cygnus). The river is sometimes identified with the river Po in Italy or with the Nile. The constellation was recorded in Ptolemy's *Almagest* in about AD 145, but probably pre-dates that – possibly back to Babylonian times (2000 BC).

Located on figures: (2.88 & 2.106), 2.69, 2.100
Area: 1140 square degrees
Number of stars visible to the naked eye under good
   conditions – 74

Named stars
Achernar – α Eri (the brightest star, magnitude 0.47)
Cursa or Kursa – β Eri
Zaurak – γ Eri
Azha – η Eri
Acamar – θ Eri (visual double star; separation 8.5″,
   magnitudes 3.4 and 4.4)

Beid – o¹ Eri
Keid – o² Eri

Double stars
θ Eri (magnitudes 3.5 and 4.5, separation 8″)
p Eri (visual binary, magnitudes 5.7 and 5.7, separation
   12″, period 500 years)

Fornax (For, Fornacis) – originally Fornax Chemica

Meaning and origin: Furnace. Introduced as a new constellation by Lacaille in his star catalogue, *Coelum Australe Stelliferum* of 1763.

Located on figures: 2.106
Area: 400 square degrees
Number of stars visible to the naked eye under good
   conditions – 11
Brightest star – α For – magnitude 3.86

Gemini (Gem, Geminorum) – a zodiacal constellation

Meaning and origin: Twins. From Greek mythology – the twin sons of Zeus (Jupiter) and Leda named Castor and Polydeuces (Pollux). When the mortal Castor was killed, the immortal Polydeuces shared his immortality, and the two are commemorated individually by the two brightest stars in the constellation (see below). The constellation was recorded in Ptolemy's *Almagest* in about AD 145, but probably pre-dates that – possibly back to Babylonian times (2000 BC).

Located on figures: 2.55, 2.62, 2.69
Area: 515 square degrees
Number of stars visible to the naked eye under good
   conditions – 45

Named stars
Castor – α Gem (visual binary, separation 4″, magnitudes
   1.9 and 2.8, period 500 years)

Fornax – Nebulae, galaxies and other objects of interest								
Object	Name	NGC	Type	RA$_{2000}$ H	m	Dec$_{2000}$ degrees	Visual mag	Size (′)
Caldwell 67	—	1097	Galaxy (S)	02	46	−30.3	9.3	9
—	—	1316	Galaxy (S)	03	23	−37.2	8.9	7

Pollux – β Gem (the brightest star, magnitude 1.15)
<u>Alhena</u> or Almeisam – γ Gem
Wasat – δ Gem
Mebsuta – ε Gem
Mekbuda – ζ Gem (variable – see below)
Propus – η Gem (variable – see below)

Variable stars
ζ Gem (Cepheid; magnitude range 4.4 to 5.2, period 10
    days)
η Gem (semi-regular variable; magnitude range 3.0 to 3.9,
    time scale 7.5 months)

Grus (Gru, Gruis)

Meaning and origin: Crane (the bird, not the lifting
device). Introduced as a new constellation by Keyser and
de Houtman around 1600.

Located on figures: <u>2.82</u>, <u>2.88</u>, 2.94, 2.111
Area: 365 square degrees
Number of stars visible to the naked eye under good
    conditions – 22

Named stars
Al Na'ir – α Gru (the brightest star, magnitude 1.73)

Variable stars
π¹ Gru (semi-regular variable, magnitude 5.5 to 6.7, time
    scale 6 months)

Hercules (Her, Herculis)

Meaning and origin: male name. From Greek mythology –
Heracles (Hercules) the fabled strong man who was set

twelve labours as a penance for killing Megara, the wife of
Eurystheus, during a fit of madness. The constellation was
recorded in Ptolemy's *Almagest* in about AD 145, but
probably pre-dates that – possibly back to Babylonian
times (2000 BC).

Located on figures: 2.37, 2.43, <u>2.48</u>, 2.94
Area: 1225 square degrees
Number of stars visible to the naked eye under good
    conditions – 80
Brightest star – ζ Her – magnitude 2.82 (visual binary,
    maximum separation 1.6″, period 34 years)

Named stars
Rasalgethi – α Her (variable – see below; visual double
    star, companion of magnitude 5.4, separation 4.6″)
Kornephoros – β Her (magnitude 2.83)

Variable stars
α Her (irregular variable, magnitude range 3.0 to 4.0, time
    scale a few months)
68 Her (eclipsing binary, magnitude range 4.7 to 5.4,
    period 2 days)

Double stars
ζ Her (visual binary; magnitudes 2.9 and 5.2, separation
    1.5″, period 34 years)

Horologium (Hor, Horologii)

Meaning and origin: Clock. Introduced as a new constellation by Lacaille in his star catalogue, *Coelum Australe
Stelliferum* of 1763.

Located on figures: (<u>2.88 & 2.100</u>)
Area: 250 square degrees

Gemini – Nebulae, galaxies and other objects of interest								
Object	Name	NGC	Type	RA₂₀₀₀ H	m	Dec₂₀₀₀ degrees	Visual mag	Size (′)
Messier 35	—	2168	Open Cluster	06	09	24.3	5.1	28
Caldwell 39	Eskimo	2392	Planetary Nebula	07	29	20.9	9.9	0.7

Hercules – Nebulae, galaxies and other objects of interest								
Object	Name	NGC	Type	RA₂₀₀₀ H	m	Dec₂₀₀₀ degrees	Visual mag	Size (′)
Messier 13	—	6205	Globular Cluster	16	42	36.5	5.9	17
—	—	6210	Planetary Nebula	16	45	23.8	9.0	14
Messier 92	—	6341	Globular Cluster	17	17	43.1	6.5	11

Number of stars visible to the naked eye under good conditions – 8

Brightest star – α Hor – magnitude 3.85

Variable stars

R Hor (long-period variable, magnitude range 4.7 to 14, period 400 days)

Hydra (Hya, Hydrae)

Meaning and origin: Sea Serpent or Water Snake. From Greek mythology – this monster haunted the source of the river Amymone in the Lernaean swamps, terrorising the local area. It had according to various accounts 8, 9, 50, 100 or even 10,000 heads, and its breath was fatal. It was slain by Heracles (Hercules) as the second of his twelve labours. A crab (Cancer – see above) attacked Heracles during the fight with the Hydra and was also killed. The constellation was recorded in Ptolemy's *Almagest* in about AD 145, but probably pre-dates that – possibly back to Babylonian times (2000 BC).

Located on figures: 2.55, 2.69, 2.76, 2.117, <u>2.122</u>

Area: 1305 square degrees (the largest of the constellations)

Number of stars visible to the naked eye under good conditions – 68

Named stars

<u>Alphard</u> or Cor Hydrae – α Hya (the brightest star, magnitude 2.0)

Variable stars

R Hya (long-period variable, magnitude range 3.5 to 11, period 1.1 years)

U Hya (semi-regular variable, magnitude range 4.2 to 6.0, time scale 1.5 years)

Hydrus (Hyi, Hydri)

Meaning and origin: Little Snake. Introduced as a new constellation by Keyser and de Houtman around 1600.

Located on figures: 2.82, <u>2.88</u>, 2.100

Area: 245 square degrees

Number of stars visible to the naked eye under good conditions – 12

Brightest star – β Hyi – magnitude 2.79

Indus (Ind, Indi)

Meaning and origin: Indian (that is, a native American). Introduced as a new constellation by Keyser and de Houtman around 1600.

Located on figures: <u>2.82</u>, <u>2.88</u>,

Area: 295 square degrees

Number of stars visible to the naked eye under good conditions – 10

Brightest star – α Ind – magnitude 3.10

Lacerta (Lac, Lacertae)

Meaning and origin: Lizard. Introduced as a new constellation by Hevelius in his star atlas, *Firmamentum Sobiescianum* of 1690.

Horologium – Nebulae, galaxies and other objects of interest								
Object	Name	NGC	Type	RA$_{2000}$ H	m	Dec$_{2000}$ degrees	Visual mag	Size (')
Caldwell 87	—	1261	Globular Cluster	03	12	–55.2	8.4	7

Hydra – Nebulae, galaxies and other objects of interest								
Object	Name	NGC	Type	RA$_{2000}$ H	m	Dec$_{2000}$ degrees	Visual mag	Size (')
Messier 48	—	2548	Open Cluster	08	14	–5.8	5.8	54
Caldwell 59	Ghost of Jupiter	3242	Planetary Nebula	10	25	–18.6	8.6	21
Messier 68	—	4590	Globular Cluster	12	40	–26.8	8.2	12
Messier 83	—	5236	Galaxy (S)	13	37	–29.9	8.2	11
Caldwell 66	—	5694	Globular Cluster	14	40	–26.5	10.2	4

Located on figures: 2.22, 2.30, 2.37
Area: 200 square degrees
Number of stars visible to the naked eye under good
    conditions – 21
Brightest star – α Lac – magnitude 3.85

Leo (Leo, Leonis) – a zodiacal constellation

Meaning and origin: Lion. From Greek mythology – the
Nemean or Cleonaean lion, which was invulnerable to
weapons made of iron, bronze or stone. It was slain by
Heracles (Hercules) as the first of his twelve labours.
Heracles choked the lion to death but lost a finger during
the fight. The constellation was recorded in Ptolemy's
Almagest in about AD 145, but probably pre-dates that –
possibly back to Babylonian times (2000 BC).

Located on figures: 2.13, 2.55, 2.122
Area: 945 square degrees
Number of stars visible to the naked eye under good
    conditions – 48

Named stars
Cor Leonis or Regulus – α Leo (brightest star, magnitude
    1.36)
Denebola – β Leo
Algeiba or Algieba – γ Leo (visual double star, separation
    4″, magnitudes 2.2 and 3.5, period 600 years)
Zosma – δ Leo
Asad Australis or Ras Elased – ε Leo

Adhafera or Aldhafera – ζ Leo
Chertan, Chort or Coxa – θ Leo
Alterf – λ Leo
Rasalas – μ Leo

Variable stars
R Leo (long-period variable, magnitude range 4.3 to 11,
    period 10 months)

Leo Minor (LMi, Leonis Minoris)

Meaning and origin: Little Lion. Introduced as a new constellation by Hevelius in his star atlas, *Firmamentum
Sobiescianum* of 1690.

Located on figures: 2.13, 2.55, 2.122
Area: 230 square degrees
Number of stars visible to the naked eye under good
    conditions – 12
Brightest star – 46 LMi – magnitude 3.81

Lepus (Lep, Leporis)

Meaning and origin: Hare. From Greek mythology – the
hare being hunted by Orion with his two dogs, Canis
Major and Canis Minor (see above). The constellation was
first recorded in Ptolemy's *Almagest* in about AD 145.

Located on figures: 2.69, 2.100, 2.106
Area: 290 square degrees
Number of stars visible to the naked eye under good
    conditions – 25

Lacerta – Nebulae, galaxies and other objects of interest							
Object	Name	NGC	Type	RA$_{2000}$ H m	Dec$_{2000}$ degrees	Visual mag	Size (′)
Caldwell 16	—	7243	Open Cluster	22 15	49.9	6.4	21

Leo – Nebulae, galaxies and other objects of interest							
Object	Name	NGC	Type	RA$_{2000}$ H m	Dec$_{2000}$ degrees	Visual mag	Size (′)
Messier 95	—	3351	Galaxy (S)	10 44	11.7	9.7	7
Messier 96	—	3368	Galaxy (S)	10 47	11.8	9.2	7
Messier 105	—	3379	Galaxy (E)	10 48	12.6	9.3	5
—	—	3521	Galaxy (S)	11 06	0.0	8.9	10
Messier 65	—	3623	Galaxy (S)	11 19	13.1	9.3	10
Messier 66	—	3627	Galaxy (S)	11 20	13.0	9.0	9
Caldwell 40	—	3626	Galaxy (S)	11 20	18.4	10.9	3

Lepus – Nebulae, galaxies and other objects of interest							
Object	Name	NGC	Type	RA₂₀₀₀ H m	Dec₂₀₀₀ degrees	Visual mag	Size (')
Messier 79	—	1904	Globular Cluster	05 25	−24.0	8.0	9

Named stars
Arneb or Arsh – α Lep (brightest star, magnitude 2.59)
Nihal – β Lep

Variable stars
μ Lep (magnetic star. magnitude range 3.0 to 3.5, time
    scale 2 days)

Libra (Lib, Librae) – a zodiacal constellation

Meaning and origin: Balance or Scales. Its name may derive from the balancing (equality) of day and night, since the Sun would have been at the autumnal equinox when in this constellation as seen in ancient Mesopotamia. The constellation was recorded in Ptolemy's *Almagest* in about AD 145, but probably pre-dates that – possibly back to Babylonian times (2000 BC).

Located on figures: 2.43, 2.48, 2.76
Area: 540 square degrees
Number of stars visible to the naked eye under good
    conditions – 32

Named stars
Zuben el genubi – α Lib
Zuben eschamali – β Lib (brightest star, magnitude 2.61)

Variable stars
δ Lib (Algol-type eclipsing binary star, magnitude range
    4.8 to 6.1, period 2.3 days)

Lupus (Lup, Lupi)

Meaning and origin: Wolf. The constellation was first recorded in Ptolemy's *Almagest* in about AD 145.

Located on figures: 2.76, 2.82
Area: 335 square degrees
Number of stars visible to the naked eye under good
    conditions – 48
Brightest star – α Lup – magnitude 2.30

Double stars
γ Lup (visual binary magnitudes 3.5 and 3.6, separation
    0.8″, period 150 years)
κ Lup (magnitudes 3.9 and 5.7, separation 27″)
π Lup (magnitudes 4.5 and 4.6, separation 1.5″)

Lynx (Lyn, Lyncis)

Meaning and origin: Lynx. Introduced as a new constellation by Hevelius in his star atlas, *Firmamentum Sobiescianum* of 1690.

Located on figures: 2.55, 2.62, 2.122
Area: 545 square degrees
Number of stars visible to the naked eye under good
    conditions – 27
Brightest star – α Lyn – magnitude 3.14

Lupus – Nebulae, galaxies and other objects of interest							
Object	Name	NGC	Type	RA₂₀₀₀ H m	Dec₂₀₀₀ degrees	Visual mag	Size (')
—	—	5822	Open Cluster	15 05	−54.4	6.5	40
—	—	5986	Globular Cluster	15 46	−37.8	7.1	10

Lynx – Nebulae, galaxies and other objects of interest							
Object	Name	NGC	Type	RA₂₀₀₀ H m	Dec₂₀₀₀ degrees	Visual mag	Size (')
Caldwell 25	—	2419	Globular Cluster	07 38	38.9	10.4	4

Lyra (Lyr, Lyrae)

Meaning and origin: Lyre. From Greek mythology – the Lyre of Orpheus. The constellation was recorded in Ptolemy's *Almagest* in about AD 145, but probably predates that – possibly back to Babylonian times (2000 BC).

Located on figures: 2.37, 2.43
Area: 285 square degrees
Number of stars visible to the naked eye under good conditions – 26

Named stars
Vega – α Lyr (the brightest star and the fifth brightest star in the night sky, magnitude 0.04; a part of the Summer Triangle)
Sheliak – β Lyr (variable – see below)
Sulafat – γ Lyr
Double-double – ε Lyr (A quadruple system. Two close *visual binaries: $ε^1$ Lyr (separation 2.6″, period 1200 years), $ε^2$ Lyr (separation 2.3″, period 600 years) are in turn separated by 208″.)

Variable stars
β Lyr (archetype of the β Lyra-type eclipsing binary stars; magnitude range 3.4 to 4.1, period 12.9 days)
R Lyr (semi-regular variable, magnitude range 3.6 to 4.9, time scale 50 days)

Double stars
ε Lyr (see above)
ζ Lyr (magnitudes 4.2 and 5.9, separation 44″)

Mensa (Men, Mensae) – originally Mons Mensae (Table Mountain)

Meaning and origin: Table. Introduced as a new constellation by Lacaille in his star catalogue, *Coelum Australe Stelliferum* of 1763. Originally named for the Table Mountain in South Africa, beneath which Lacaille made his observations.

Located on figures: 2.88, 2.100
Area: 155 square degrees

Number of stars visible to the naked eye under good conditions – 6
Brightest star – α Men – magnitude 5.08. The most inconspicuous of the constellations, and the only one not to include a single fourth-magnitude or brighter star, though it does share the Large Magellanic Cloud with Dorado.

Microscopium (Mic, Microscopii)

Meaning and origin: Microscope. Introduced as a new constellation by Lacaille in his star catalogue, *Coelum Australe Stelliferum* of 1763.

Located on figures: 2.82, 2.88, 2.94
Area: 210 square degrees
Number of stars visible to the naked eye under good conditions – 10
Brightest star – γ Mic – magnitude 4.66

Monoceros (Mon, Monocerotis)

Meaning and origin: Unicorn. Introduced as a new constellation by Plancius on his star globe of 1613.

Located on figures: 2.55, 2.69, 2.100, 2.117
Area: 480 square degrees
Number of stars visible to the naked eye under good conditions – 32
Brightest star – β Mon – magnitude 3.8 (a visual triple star, magnitudes 4.6, 5.2 and 5.6, separations 7.4″ and 2.8″)

Double stars
β Mon (magnitudes 4.7 and 5.2, separation 7″)
ε Mon (magnitudes 4.5 and 6.5, separation 14″)

Musca (Mus, Muscae) – originally Musca Australis (Southern Fly)

Meaning and origin: Fly. Introduced as a new constellation by Keyser and de Houtman around 1600.

Located on figures: 2.76, 2.100
Area: 140 square degrees

Object	Name	NGC	Type	RA₂₀₀₀ H	m	Dec₂₀₀₀ degrees	Visual mag	Size (′)

Lyra – Nebulae, galaxies and other objects of interest

Object	Name	NGC	Type	RA$_{2000}$ H	m	Dec$_{2000}$ degrees	Visual mag	Size (′)
Messier 57	Ring	6720	Planetary Nebula	18	54	33.0	9.7	3
Messier 56	—	6779	Globular Cluster	19	17	30.2	8.3	7

Monoceros – Nebulae, galaxies and other objects of interest

Object	Name	NGC	Type	RA2000 H	m	Dec2000 degrees	Visual mag	Size (')
Caldwell 50*	—	2244	Open Cluster	06	32	4.9	4.8	24
Caldwell 49	Rosette	2237–9	Emission Nebula	06	32	5.1	≈ 4	80
Caldwell 46	Hubble's variable	2261	Emission Nebula	06	39	8.7	10.0	2
—	—	2286	Open Cluster	06	48	–3.2	7.5	15
—	—	2301	Open Cluster	06	52	0.5	6.0	12
Messier 50	—	2323	Open Cluster	07	03	–8.3	5.9	16
Caldwell 54	—	2506	Open Cluster	08	00	–10.8	7.6	7

* Inside Caldwell 49.

Musca – Nebulae, galaxies and other objects of interest

Object	Name	NGC	Type	RA2000 H	m	Dec2000 degrees	Visual mag	Size (')
Caldwell 108	—	4372	Globular Cluster	12	26	–72.7	7.8	19
Caldwell 105	—	4833	Globular Cluster	13	00	–70.9	7.4	14

Number of stars visible to the naked eye under good conditions – 14

Brightest star – α Mus – magnitude 2.71

Double stars

β Mus (visual binary, magnitudes 3.7 and 4.1, separated by 1.5″, period 400 years)

Norma (Nor, Normae)

Meaning and origin: Rule (that is, a straight-edge, not a regulation). Introduced as a new constellation by Lacaille in his star catalogue, *Coelum Australe Stelliferum* of 1763.

Located on figures: 2.76, 2.82
Area: 165 square degrees
Number of stars visible to the naked eye under good conditions – 10
Brightest star – γ² Nor – magnitude 4.01

Octans (Oct, Octantis) – originally Octans Hadleianus (Hadley's Octant)

Meaning and origin: Octant. Introduced as a new constellation by Lacaille in his star catalogue, *Coelum Australe Stelliferum* of 1763.

Located on figures: 2.76, 2.82, 2.88
Area: 290 square degrees
The celestial South Pole is contained within this constellation, and the star σ Oct (magnitude 5.46) is the nearest naked-eye star to the pole, though it is far too faint to be classed as a useful pole star.
Number of stars visible to the naked eye under good conditions – 14
Brightest star – ν Oct – magnitude 3.75

Variable stars

ε Oct (semi-regular variable, magnitude range 4.5 to 5.5, time scale 2 months)

Norma – Nebulae, galaxies and other objects of interest

Object	Name	NGC	Type	RA2000 H	m	Dec2000 degrees	Visual mag	Size (')
Caldwell 89	S Norma	6087	Open Cluster	16	19	–57.9	5.4	12

Ophiuchus (Oph, Ophiuchi) – an "unofficial" zodiacal constellation

Meaning and origin: Serpent Carrier. From Greek mythology – also known as Asclepius (Aesculapius), Ophiuchus was a highly skilled physician able to restore life to the dead. Hades (Pluto), lord of the underworld, complained to Zeus (Jupiter) at the diminution in the numbers of his subjects occurring through Ophiuchus' healing powers, and so Ophiuchus was killed by a thunderbolt. Zeus later restored Ophiuchus to life, and after his second demise, his image, bearing a curative serpent, was set among the stars by Zeus. The constellation was first recorded in Ptolemy's *Almagest* in about AD 145.

Located on figures: 2.43, 2.48, 2.82, 2.94
Area: 950 square degrees
Ophiuchus, the serpent carrier, separates the two halves of Serpens (Serpens Caput – the head of the serpent, and Serpens Cauda, the body of the serpent).
Number of stars visible to the naked eye under good conditions – 51

Named stars
Ras alague or Ras alhague – α Oph (the brightest star, magnitude 2.08)
Cebalrai or Cheleb – β Oph
Yed Prior – δ Oph
Yed Posterior – ε Oph
Sabik – η Oph
Marfik – λ Oph (visual binary; magnitudes 4.2 and 5.2, separation 1.5″, period 130 years)

Variable stars
κ Oph (irregular variable, magnitude range 2.8 to 3.6)
χ Oph (emission line star, magnitude range 4.2 to 5.1)

RS Oph (recurrent nova, magnitude range 4 to 12, interval between outbursts 10 to 30 years)

Double stars
70 Oph (visual binary; magnitudes 4.2 and 6.0, separation 1.5″ to 7″, period 88 years)

Orion (Ori, Orionis)

Meaning and origin: male name. From Greek mythology – a hunter in Boeotia, and the handsomest man of his time, Orion sought to marry Merope. He rapidly succeeded in the nuptial task of ridding the island of Chios, which was given to him by Merope's father, Oenopion, of wild beasts in order to be worthy of her hand in marriage. Oenopion, however, then refused to honour the bargain, and a drunken Orion ravished Merope. Oenopion plied Orion with further wine until he was comatose, and then blinded him. Orion's sight was restored by Helius, and he sought revenge upon Oenopion. Apollo, however, had been angered by boasts made by Orion of his prowess as a hunter and arranged for him to be attacked by an invulnerable scorpion. Orion fled the scorpion by swimming out to sea, where he was shot and killed in error by an arrow from Artemis (Diana). In atonement, Artemis set Orion's image amongst the stars, opposite to the scorpion (Scorpio) in the sky. (Some versions of the legend have Orion being killed by the scorpion.) The constellation was first recorded in Ptolemy's *Almagest* in about AD 145.

Located on figures: 2.69, 2.100, 2.106
Area: 595 square degrees
Number of stars visible to the naked eye under good conditions – 70

Named stars
Betelgeuse – α Ori (variable star – see below)

Ophiuchus – Nebulae, galaxies and other objects of interest							
Object	Name	NGC	Type	RA₂₀₀₀ H m	Dec₂₀₀₀ degrees	Visual mag	Size (′)
Messier 107	—	6171	Globular Cluster	16 33	−13.1	8.1	10
Messier 12	—	6218	Globular Cluster	16 47	−2.0	6.6	15
Messier 10	—	6254	Globular Cluster	16 57	−4.1	6.6	15
Messier 62	—	6266	Globular Cluster	17 01	−30.1	6.6	14
Messier 19	—	6273	Globular Cluster	17 03	−26.3	7.2	14
Messier 9	—	6333	Globular Cluster	17 19	−18.5	7.9	9
Messier 14	—	6402	Globular Cluster	17 38	−3.3	7.6	12
—	—	IC 4665	Open Cluster	17 46	5.7	4.2	40
—	—	6633	Open Cluster	18 28	6.6	4.6	27

Rigel – β Ori (the brightest star, magnitude 0.08; a visual double with a sixth-magnitude companion 9″ away)

Bellatrix – γ Ori

Mintaka – δ Ori

Alnilam – ε Ori ⎫
Alnitak – ζ Ori (double star, magnitudes 1.9 and 4.1, separation 2.5″) ⎬ Orion's belt

Trapezium – θ Ori (A multiple star system at the heart of the Orion nebula (M42 – see below). It contains four or more stars with separations ranging from 9″ to 19″, and magnitudes ranging from 5.4 to 7 (variable). The stars form the corners of a distorted rectangle, from which shape arises the name.)

Saiph – κ Ori

Meissa – λ Ori

Variable stars

Betelgeuse (α Ori – semi-regular variable, magnitude range 0.1 to 1.3, period 6 years)

U Ori (long-period variable, magnitude range 4.5 to 13, period 1 year)

Double stars

ζ Ori (visual binary, magnitudes 1.9 and 4.2, separation 2.5″, period 1500 years)

η Ori (magnitudes 3.9 and 4.7, separation 1.5″)

λ Ori (magnitudes 3.4 and 5.5, separation 4.5″)

Pavo (Pav, Pavonis)

Meaning and origin: Peacock. Introduced as a new constellation by Keyser and de Houtman around 1600.

Located on figures: 2.82, 2.88
Area: 380 square degrees
Number of stars visible to the naked eye under good conditions – 25

Named stars

Peacock – α Pav (the brightest star, magnitude, 1.93)

Variable stars

κ Pav (Cepheid variable, magnitude range 3.9 to 4.8, period 9 days)

λ Pav (Shell star, magnitude range 3.4 to 4.3)

Pegasus (Peg, Pegasi)

Meaning and origin: Flying Horse. From Greek mythology – conceived of Poseidon and Medusa, Pegasus was released from the Gorgon's dead body after she had been decapitated by Perseus. Later Bellerophon rode Pegasus to attack and kill the Chimaera, a fire-breathing monster with the tail of a snake, the body of a goat, and the head of a lion. Bellerophon then attempted to fly Pegasus to Olympus, but Zeus (Jupiter) caused a horse fly to bite Pegasus so that the horse bucked and flung Bellerophon off to fall back to Earth. Pegasus however continued the journey to Olympus, and duly took his place in the sky. The constellation was recorded in Ptolemy's *Almagest* in about AD 145, but probably pre-dates that – possibly back to Babylonian times (2000 BC).

Located on figures: 2.24, 2.30, 2.37, 2.94, 2.111
Area: 1120 square degrees.

Orion – Nebulae, galaxies and other objects of interest								
Object	Name	NGC	Type	RA₂₀₀₀ H	m	Dec₂₀₀₀ degrees	Visual mag	Size (′)
—	—	1981	Open Cluster	05	35	−4.4	4.6	25
Messier 42	Orion	1976	Emission Nebula	05	35	−5.5	4.0	66
Messier 43	Orion	1982	Emission Nebula	05	36	−5.3	9.0	20
Messier 78	—	2068	Reflection Nebula	05	47	0.1	8.0	8
—	—	2175	Open Cluster	06	10	20.3	6.8	18

Pavo – Nebulae, galaxies and other objects of interest								
Object	Name	NGC	Type	RA₂₀₀₀ H	m	Dec₂₀₀₀ degrees	Visual mag	Size (′)
Caldwell 101	—	6744	Galaxy (S)	19	10	−63.9	8.4	16
Caldwell 93	—	6752	Globular Cluster	19	11	−60.0	5.4	20

Number of stars visible to the naked eye under good conditions – 52

Named stars

Marchab or <u>Markab</u> – α Peg (magnitude 2.56; same name as κ Vel)

Scheat – β Peg (variable – see below)

Algenib – γ Peg (also the name of α Per)

Enif – ε Peg (the brightest star; irregular variable, magnitude normally about 2.4)

Homam – ζ Peg

Matar – η Peg

Biham – θ Peg

Sadalbari – μ Peg

Variable stars

β Peg (semi-regular variable, magnitude range 2.4 to 2.8, time scale 1 month)

Perseus (Per, Persei)

Meaning and origin; male name. From Greek mythology – Perseus undertook to slay Medusa in order to prevent an unwanted marriage being forced upon his mother Danaë. He obtained winged sandals, a magic bag to contain Medusa's head and a helmet of invisibility from the Stygian Nymphs, a brightly polished shield from Athene (Minerva), and a diamond sickle from Hermes (Mercury). He avoided the petrifying look from Medusa, by watching only her reflection in the shield, and cut off her head with the sickle. Pegasus and the warrior Chrysaor sprang from the dead body fully grown. The invisibility helmet enabled Perseus to escape Medusa's sisters, Stheno and Euryale. On his return journey, Perseus rescued Andromeda from the Hydra by decapitating the latter with the sickle. Perseus was promised Andromeda as his wife by her parents Cepheus and Cassiopeia as a reward for the rescue. When they tried to back down from the arrangement afterwards, Perseus used Medusa's head to turn his opponents to stone. In the same fashion he rescued his mother from her unwanted suitor Polydectes. Later Perseus was to kill his own grandfather with a discus while competing in funeral games. The constellation was recorded in Ptolemy's *Almagest* in about AD 145, but probably pre-dates that – possibly back to Babylonian times (2000 BC).

Located on figures: 2.22, <u>2.24</u>, <u>2.62</u>
Area: 615 square degrees
Number of stars visible to the naked eye under good conditions – 60

Named stars

Algenib, Marfak or <u>Mirfak</u> – α Per (the brightest star, magnitude 1.79; the first name is shared with γ Peg)

Algol – β Per (variable – see below)

Pegasus – Nebulae, galaxies and other objects of interest				RA$_{2000}$		Dec$_{2000}$	Visual	Size
Object	Name	NGC	Type	H	m	degrees	mag	(')
Caldwell 43	—	7814	Galaxy (S)	00	03	16.2	10.5	6
Messier 15	—	7078	Globular Cluster	21	30	12.2	6.4	12
Caldwell 30	—	7331	Galaxy (S)	22	37	34.4	9.5	11
Caldwell 44	—	7479	Galaxy (S)	23	05	12.3	11.0	4

Perseus – Nebulae, galaxies and other objects of interest				RA$_{2000}$		Dec$_{2000}$	Visual	Size
Object	Name	NGC	Type	H	m	degrees	mag	(')
Messier 76	Little Dumbbell	650	Planetary Nebula	01	42	51.6	12.2	5
Caldwell 14	h & χ Per	869/884	Open Cluster	02	20	57.1	4.3/4.4	30/30
—	—	1023	Galaxy (E)	02	40	39.1	9.3	9
Messier 34	—	1039	Open Cluster	02	42	42.8	5.2	35
—	—	1245	Open Cluster	03	15	47.3	8.4	10
Caldwell 24	—	1275	Galaxy (I)	03	20	41.5	11.6	3
—	—	1342	Open Cluster	03	32	37.3	6.7	14
—	California	1499	Emission Nebula	04	01	51.3	7	140
—	—	1528	Open Cluster	04	15	51.2	6.4	24

Atik – ζ Per
Menkib – ξ Per

Variable stars
β Per (the archetypal eclipsing binary star; magnitude
  range 2.1 to 3.4, period 2.9 days)
ρ Per (semi-regular variable; magnitude range 3.2 to 4.1,
  time scale 2 months)

Phoenix (Phe, Phoenicis)

Meaning and origin: Phoenix. A legendary bird, wor-
shipped in Egypt, that was consumed by fire every 500
years, but then rose rejuvenated from the ashes.
Introduced as a new constellation by Keyser and de
Houtman around 1600.

Located on figures: 2.88, 2.111
Area: 470 square degrees.
Number of stars visible to the naked eye under good
  conditions – 20

Named stars
Ankaa – α Phe (the brightest star, magnitude 2.39)

Variable stars
ζ Phe (eclipsing binary, magnitude range 4.0 to 4.5, period
  40 hours)

Double stars
β Phe (visual binary, magnitudes 4.0 and 4.2, separation
  1.5″)

Pictor (Pic, Pictoris) – originally Equuleus Pictoris (the Painter's Easel)

Meaning and origin: Painter. Introduced as a new constellation by Lacaille in his star catalogue, *Coelum Australe
Stelliferum* of 1763.

Located on figures: 2.100
Area: 245 square degrees.
Number of stars visible to the naked eye under good
  conditions – 10
Brightest star – α Pic – magnitude 3.26

Pisces (Psc, Piscium) – a zodical constellation

Meaning and origin: Fishes. From Greek mythology –
when the giant Typhon attacked Olympus the gods fled to
Egypt and hid in the form of various animals. Aphrodite
(Venus) and Eros (Cupid) transformed themselves into
fish and hid in the river Euphrates. This constellation
commemorates that episode. The constellation was
recorded in Ptolemy's *Almagest* in about AD 145, but
probably pre-dates that – possibly back to Babylonian
times (2000 BC).

Located on figures: 2.24, 2.30, 2.106, 2.111
Area: 890 square degrees
Number of stars visible to the naked eye under good
  conditions – 45
Brightest star – η Psc – magnitude 3.62

Named stars
Al rescha – α Psc (visual binary, magnitudes 4.3 and 5.2,
  separation ranges from 1″ to 4″, period 900 years)

Variable stars
TV Psc (semi-regular variable; magnitude range 4.7 to 5.5,
  time scale 50 days)
TX Psc (irregular variable; magnitude range 4.6 to 5.2)

Double stars
ψ¹ Psc (magnitudes 5.6 and 5.8, separation 30″)

Piscis Austrinus (PsA, Piscis Austrini)

Meaning and origin: Southern Fish. The constellation was
recorded in Ptolemy's *Almagest* in about AD 145, but
probably pre-dates that – possibly back to Babylonian
times (2000 BC).

Located on figures: 2.88, 2.94, 2.111
Area: 245 square degrees.
Number of stars visible to the naked eye under good
  conditions – 12

Named stars
Fomalhaut – α PsA (the brightest star, magnitude 1.16)

Pisces – Nebulae, galaxies and other objects of interest							
Object	Name	NGC	Type	RA₂₀₀₀ H m	Dec₂₀₀₀ degrees	Visual mag	Size (′)
Messier 74	—	628	Galaxy (S)	01 37	15.8	9.2	10

Puppis (Pup, Puppis)

Meaning and origin: Poop or Stern. Originally this constellation along with Carina (keel) and Vela (sails), formed the constellation Argo (ship – also called Argo Navis, and one of Ptolemy's constellations). Argo, in Greek mythology, was Jason's ship as used by the argonauts during the hunt for the golden fleece. In Lacaille's star catalogue, *Coelum Australe Stelliferum* of 1763, Argo was subdivided into the three smaller constellations for convenience.

Located on figures: 2.100, 2.117
Area: 675 square degrees
Number of stars visible to the naked eye under good conditions – 84

Named stars
Naos or Suhail Hadar – ζ Pup (the brightest star, magnitude 2.2)

Variable stars
V Pup (eclipsing binary, magnitude range 4.4 to 4.8, period 35 hours)
L^2 Pup (semi-regular variable, magnitude range 2.5 to 6.0, time scale 4 months)

Double stars
k Pup (magnitudes 4.5 and 4.8, separation 10″)

Pyxis (Pyx, Pyxidis) – originally Malus, then Pyxis Nautica

Meaning and origin: Mariner's Compass. Introduced as a new constellation by Lacaille in his star catalogue, *Coelum Australe Stelliferum* of 1763.

Located on figures: 2.100, 2.117
Area: 220 square degrees

Number of stars visible to the naked eye under good conditions – 10
Brightest star – α Pyx – magnitude 3.68

Reticulum (Ret, Reticuli) – originally Reticulum Rhomboidalis (Rhomboidal Net)

Meaning and origin: Net. The name derives from the network of cross-wires used by Lacaille inside the eyepiece of his telescope to determine star positions (compare the modern usage: reticle). Introduced as a new constellation by Lacaille in his star catalogue, *Coelum Australe Stelliferum* of 1763.

Located on figures: 2.88, 2.100
Area: 115 square degrees
Number of stars visible to the naked eye under good conditions – 8
Brightest star – α Ret – magnitude 3.34

Sagitta (Sge, Sagittae)

Meaning and origin: Arrow. Possibly Eros' (Cupid's) arrow, or the arrow that killed Orion (see above), or an arrow from the centaur Cheiron (see Sagittarius) aimed at the scorpion (Scorpio), or the arrow shot by Apollo to kill the Cyclops, or the arrow shot by Heracles (Hercules) to kill the griffon–vulture that was tormenting Prometheus. The constellation was first recorded in Ptolemy's *Almagest* in about AD 145.

Located on figures: 2.37, 2.43, 2.94
Area: 80 square degrees
Number of stars visible to the naked eye under good conditions – 6
Brightest star – γ Sge – magnitude 3.56

Object	Name	NGC	Type	RA$_{2000}$ H	m	Dec$_{2000}$ degrees	Visual mag	Size (′)
Messier 47	—	2422	Open Cluster	07	37	−14.5	4.4	30
Messier 46	—	2437	Open Cluster	07	42	−14.8	6.1	27
Messier 93	—	2447	Open Cluster	07	45	−23.9	6.2	22
—	—	2451	Open Cluster	07	45	−38.0	2.8	45
Caldwell 71	—	2477	Open Cluster	07	52	−38.6	5.8	27
—	—	2527	Open Cluster	08	05	−28.2	6.5	22

Puppis – Nebulae, galaxies and other objects of interest

Sagitta – Nebulae, galaxies and other objects of interest							
Object	Name	NGC	Type	RA2000 H m	Dec2000 degrees	Visual mag	Size (')
Messier 71	—	6838	Globular Cluster	19 54	18.8	8.3	7

Sagittarius (Sgr, Sagittarii) – a zodiacal constellation

Meaning and origin: Archer. From Greek mythology – represents Cheiron, the wise, learned and immortal king of the centaurs, who acted as tutor to Asclepius (see notes on Ophiuchus), Jason and Heracles (Hercules). Cheiron was wounded in the knee by a poisoned arrow shot by Heracles during his fourth labour. To escape the agony of his wound, Cheiron relinquished his immortality to Prometheus enabling the latter to be released from a captivity in which a griffon–vulture tore at his liver daily. An alternative legend has it that the constellation represents Crotus who invented the bow and arrow. The constellation was recorded in Ptolemy's *Almagest* in about AD 145 but probably pre-dates that – possibly back to Babylonian times (2000 BC).

Located on figures: 2.43, 2.82, 2.94
Area: 865 square degrees.
The centre of our own galaxy, the Milky Way, is in Sagittarius about 3° west of γ Sgr
Number of stars visible to the naked eye under good conditions – 60

Named stars
Rukbat or Alrami – α Sgr
Arkab – β Sgr
Alnasl – γ Sgr
Kaus Media – δ Sgr
Kaus Australis – ε Sgr (the brightest star, magnitude 1.84)
Ascella – ζ Sgr
Al kitn – η Sgr
Kaus Borealis – λ Sgr
Nunki – σ Sgr

Variable stars
W Sgr (Cepheid, magnitude range 4.3 to 5.0, period 7.5 days)
X Sgr (Cepheid, magnitude range 4.0 to 4.8, period 7 days)

Scorpius (Sco, Scorpii) – also known as Scorpio – a zodiacal constellation

Meaning and origin: Scorpion. From Greek mythology – this is the immortal scorpion from which Orion (see above) fled to his death. The constellation was recorded in Ptolemy's *Almagest* in about AD 145, but probably pre-dates that – possibly back to Babylonian times (2000 BC).

Sagittarius – Nebulae, galaxies and other objects of interest							
Object	Name	NGC	Type	RA2000 H m	Dec2000 degrees	Visual mag	Size (')
Messier 23	—	6494	Open Cluster	17 57	−19.0	5.5	27
Messier 20	Trifid	6514	Emission Nebula	18 03	−23.0	6.3	29
Messier 8	Lagoon	6523	Emission Nebula	18 04	−24.4	5.8	90
Messier 21	—	6531	Open Cluster	18 05	−22.5	5.9	13
Messier 24	—	6603	Star Cloud	18 18	−18.4	11.1	5
Messier 18	—	6613	Open Cluster	18 20	−17.1	6.9	9
Messier 17	Omega	6618	Emission Nebula	18 21	−16.2	6.0	46
Messier 28	—	6626	Globular Cluster	18 25	−24.9	6.9	11
Messier 69	—	6637	Globular Cluster	18 31	−32.4	7.7	7
Messier 25	—	IC 4725	Open Cluster	18 32	−19.3	4.6	32
Messier 22	—	6656	Globular Cluster	18 36	−23.9	5.1	24
Messier 70	—	6681	Globular Cluster	18 43	−32.3	8.8	8
Messier 54	—	6715	Globular Cluster	18 55	−30.5	7.7	9
Messier 55	—	6809	Globular Cluster	19 40	−31.0	7.0	19
Caldwell 57	Barnard's	6822	Galaxy (I)	19 45	−14.8	9.4	10
Messier 75	—	6864	Globular Cluster	20 06	−21.9	8.6	6

Located on figures: 2.43, 2.48, 2.76, 2.82, 2.94
Area: 495 square degrees
Number of stars visible to the naked eye under good
conditions – 55

Named stars
Antares – α Sco (The brightest star, irregular variable, mag-
nitude range 0.9 to 1.1. The name derives from the star's
red colour and means "Rival of Mars". It may easily be
mistaken for the planet by the inexperienced observer.)
Akrab or Graffias – β Sco (visual double star, magnitudes
2.6 and 4.8, separation 14″)
Dschubba – δ Sco
Shaula – λ Sco
Lesath – ν Sco

Double stars
ξ Sco (visual binary, magnitudes 4.8 and 4.8, separation
0.5″, period 45 years)

Sculptor (Scl, Sculptoris) – originally Apparatus Sculptoris (Sculptor's equipment)

Meaning and origin: Sculptor. Introduced as a new constellation by Lacaille in his star catalogue, *Coelum
Australe Stelliferum* of 1763.

Located on figures: 2.88, 2.106, 2.111
Area: 475 square degrees
Number of stars visible to the naked eye under good
conditions – 12
Brightest star – α Scl – magnitude 4.30

Scutum (Sct, Scuti) – originally Scutum Sobieskii

Meaning and origin: Shield of Arms. From the shield of
arms of Johannes Sobieski III, king of Poland. Introduced
as a new constellation by Hevelius in his star atlas,
Firmamentum Sobiescianum of 1690.

Located on figures: 2.43, 2.94
Area: 110 square degrees
Number of stars visible to the naked eye under good
conditions – 7
Brightest star – α Sct – magnitude 3.84

Variable stars
δ Sct (archetype of the δ Scuti variables, magnitude range
4.9 to 5.2, period 4.5 hours)
R Sct (RV Tau variable, magnitude range 4.9 to 8.0, period
about 145 days)

Scorpius – Nebulae, galaxies and other objects of interest

Object	Name	NGC	Type	RA2000 H	m	Dec2000 degrees	Visual mag	Size (′)
Messier 80	—	6093	Globular Cluster	16	17	–23.0	7.2	9
Messier 4	—	6121	Globular Cluster	16	24	–26.5	5.9	26
Caldwell 75	—	6124	Open Cluster	16	26	–40.7	5.8	29
Caldwell 76	—	6231	Open Cluster	16	54	–41.8	2.6	15
Caldwell 69	Bug	6302	Planetary Nebula	17	14	–37.1	12.8	1
—	—	6388	Globular Cluster	17	36	–44.7	6.9	9
Messier 6	Butterfly	6405	Open Cluster	17	40	–32.2	4.2	15
—	—	6416	Open Cluster	17	44	–32.4	5.7	18
Messier 7	—	6475	Open Cluster	17	54	–34.8	3.3	80

Sculptor – Nebulae, galaxies and other objects of interest

Object	Name	NGC	Type	RA2000 H	m	Dec2000 degrees	Visual mag	Size (′)
—	—	7793	Galaxy (S)	23	58	–32.6	9.0	9
Caldwell 72	—	55	Galaxy (S)	00	15	–39.2	7.9	32
Caldwell 65	Silver Coin	253	Galaxy (S)	00	48	–25.3	7.1	25
—	—	288	Globular Cluster	00	53	–26.6	8.1	14
Caldwell 70	—	300	Galaxy (S)	00	55	–37.7	8.7	20

Scutum – Nebulae, galaxies and other objects of interest							
Object	Name	NGC	Type	RA2000 H m	Dec2000 degrees	Visual mag	Size (')
Messier 26	—	6694	Open Cluster	18 45	−9.4	8.0	15
Messier 11	Wild Duck	6705	Open Cluster	18 51	−6.3	5.8	14

Serpens (Ser, Serpentis)

Meaning and origin: Serpent. From Greek mythology – the serpent carried by Ophiuchus (see above). This is a serpent of healing in Ophiuchus' guise as the physician Asclepius (Aesculapius). But other versions of the legend have the two involved in a death struggle. The constellation was recorded in Ptolemy's *Almagest* in about AD 145, but probably pre-dates that – possibly back to Babylonian times (2000 BC).

This constellation is unique, being in two sections. The sections are Serpens Caput (the head of the serpent), and Serpens Cauda (the body or tail of the serpent), and they are separated by Ophiuchus, the serpent carrier.

Located on figures: <u>2.43</u>, 2.48, 2.94
Total area: 635 square degrees
Number of stars visible to the naked eye under good
conditions – 30

Named stars
Unukalhai – α Ser (the brightest star, magnitude 2.65)
Alya – θ Ser

Double stars
δ Ser (magnitudes 4.2 and 5.2, separation 4″)
θ Ser (magnitudes 4.5 and 5.4, separation 22″)

Sextans (Sex, Sextantis) – originally Sextans Uraniæ (Heavenly Sextant)

Meaning and origin: Sextant. Introduced as a new constellation by Hevelius in his star atlas, *Firmamentum Sobiescianum* of 1690.

Located on figures: <u>2.55</u>, <u>2.122</u>
Area: 315 square degrees
Number of stars visible to the naked eye under good
conditions – 4
Brightest star – α Sex – magnitude 4.49

Taurus (Tau, Tauri) – a zodiacal constellation

Meaning and origin: Bull. From Greek mythology – Zeus (Jupiter) took the form of a white bull in order to lure Europa away. With Europa on his back he swam from Canaan to Crete. Once there he transformed into an eagle in order to consummate their union. The constellation was recorded in Ptolemy's *Almagest* in about AD 145 but probably pre-dates that – possibly back to Babylonian times (2000 BC).

Located on figures: 2.24, 2.62, <u>2.69</u>, 2.106

Serpens – Nebulae, galaxies and other objects of interest							
Object	Name	NGC	Type	RA2000 H m	Dec2000 degrees	Visual mag	Size (')
Messier 16	—	6611	Open Cluster*	18 19	−13.8	6.0	7
—	Eagle	IC 4703	Emission Nebula*	18 19	−13.8	8	35
Messier 5	—	5904	Globular Cluster	15 19	2.1	5.8	17

* The star cluster is embedded inside the nebula, so that these two are really one object.

Sextans – Nebulae, galaxies and other objects of interest							
Object	Name	NGC	Type	RA2000 H m	Dec2000 degrees	Visual mag	Size (')
Caldwell 53	Spindle	3115	Galaxy (E)	10 05	−7.7	9.2	8

Area: 795 square degrees
Number of stars visible to the naked eye under good
   conditions – 90

Named stars
Aldebaran – α Tau (the brightest star, magnitude 0.86)
El Nath – β Tau
Celaeno – 16 Tau ⎫
Electra – 17 Tau
Taygeta – 19 Tau
Maia – 20 Tau
Asterope or Sterope – 21 Tau ⎬ The Pleiades
Merope – 23 Tau
Alcyone – η Tau
Atlas – 27 Tau
Pleione – 28 Tau (variable star – see below) ⎭

Variable stars
λ Tau (eclipsing binary, magnitude range 3.4 to 3.9,
   period 4 days)
Pleione (28 Tau or BU Tau, Shell star; magnitude range
   4.6 to 5.5)

Double stars
θ Tau (magnitudes 3.4 and 3.8, separation 337″)

Telescopium (Tel, Telescopii)

Meaning and origin: Telescope. Introduced as a new
constellation by Lacaille in his star catalogue, *Coelum
Australe Stelliferum* of 1763.

Located on figures: 2.82
Area: 250 square degrees

Number of stars visible to the naked eye under good
   conditions – 12
Brightest star – α Tel – magnitude 3.50

Triangulum (Tri, Trianguli)

Meaning and origin: Triangle. The constellation was first
recorded in Ptolemy's *Almagest* in about AD 145.

Located on figures: 2.24, 2.30, 2.62, 2.111
Area: 130 square degrees
Number of stars visible to the naked eye under good
   conditions – 10
Brightest star – β Tri – magnitude 3.00

Named stars
Caput Trianguli – α Tri

Triangulum Australe (TrA, Trianguli Australis)

Meaning and origin: Southern Triangle. Introduced as a
new constellation by Keyser and de Houtman around
1600.

Located on figures: 2.76, 2.82, 2.88
Area: 110 square degrees
Number of stars visible to the naked eye under good
   conditions – 10

Named stars
Atria – α TrA (the brightest star, magnitude 1.91)

Taurus – Nebulae, galaxies and other objects of interest

Object	Name	NGC	Type	RA$_{2000}$ H	m	Dec$_{2000}$ degrees	Visual mag	Size (′)
Messier 45	Pleiades		Open Cluster	03	47	24.1	1.2	110
Caldwell 41	Hyades		Open Cluster	04	27	16.0	0.5	330
—	—	1647	Open Cluster	04	46	19.1	6.4	45
Messier 1	Crab	1952	Supernova Remnant	05	35	22.0	8.4	6

Triangulum – Nebulae, galaxies and other objects of interest

Object	Name	NGC	Type	RA$_{2000}$ H	m	Dec$_{2000}$ degrees	Visual mag	Size (′)
Messier 33	Triangulum	598	Galaxy (S)	01	34	30.7	5.7	62

Triangulum Australe – Nebulae, galaxies and other objects of interest								
Object	Name	NGC	Type	RA₂₀₀₀ H	m	Dec₂₀₀₀ degrees	Visual mag	Size (')
Caldwell 95	—	6025	Open Cluster	16	04	−60.5	5.1	12

Tucana (Tuc, Tucanae)

Meaning and origin: Toucan. Introduced as a new constellation by Keyser and de Houtman around 1600.

Located on figures: 2.82, 2.88
Area: 295 square degrees
Number of stars visible to the naked eye under good conditions – 12
Brightest star – α Tuc – magnitude 1.39

Double stars
β Tuc (quadruple system, two close double stars, magnitudes 4.3 and 4.4, separated by 27″)

Ursa Major (UMa, Ursae Majoris)

Meaning and origin: Great Bear. From Greek mythology – Callisto, a handmaiden of Artemis (Diana), was seduced by Zeus (Jupiter). For punishment, Artemis (goddess of chastity as well as of hunting) transformed Callisto into a bear, and set her pack of dogs to hunt the bear to death. Zeus saved Callisto by placing her, still in bear-form, in the sky. Her son by Zeus, Arcas, joined her as Ursa Minor. The constellation was first recorded in Ptolemy's *Almagest* in about AD 145.

Located on figures: 2.13, 2.22, 2.55, 2.62
Area: 1280 square degrees
Number of stars visible to the naked eye under good conditions – 65

Named stars
Dubhe – α UMa
Merak – β UMa
Phecda – γ UMa
Megrez – δ UMa
Alioth – ε UMa (the brightest star, magnitude 1.76)
Mizar – ζ UMa (also the name of ε Boö)
Alcaid, Alkaid, Benatnasch or Benetnasch – η UMa

The stars of the Plough or Big Dipper

Talitha – ι UMa
Tania Borealis – λ UMa
Tania Australis – μ UMa
Alula Borealis – ν UMa

Tucana – Nebulae, galaxies and other objects of interest								
Object	Name	NGC	Type	RA₂₀₀₀ H	m	Dec₂₀₀₀ degrees	Visual mag	Size (')
Small Magellanic Cloud	—	—	Galaxy (I)	00	55	−73.0	2.4	250
Caldwell 106	47 Tucanae	104	Globular Cluster	00	24	−72.1	4.0	31
Caldwell 104	—	362	Globular Cluster	01	03	−70.9	6.6	13

Ursa Major – Nebulae, galaxies and other objects of interest								
Object	Name	NGC	Type	RA₂₀₀₀ H	m	Dec₂₀₀₀ degrees	Visual mag	Size (')
Messier 81	Bode's	3031	Galaxy (S)	09	56	69.1	6.9	26
Messier 82	—	3034	Galaxy (I)	09	56	69.7	8.4	11
Messier 108	—	3556	Galaxy (S)	11	12	55.7	10.1	8
Messier 97	Owl	3587	Planetary Nebula	11	15	55.0	12.0	3
Messier 109	—	3992	Galaxy (S)	11	58	53.4	9.8	8
Messier 40	—	—	2 stars	12	22	58.1	0.8	9
Messier 101	Pinwheel	5457	Galaxy (S)	14	03	54.4	7.7	27

Alula Australis – ξ UMa (visual binary, magnitudes 4.3 and 4.8, separation 2″, period 60 years)
Muscida – o UMa
Alcor – 80 UMa

Double stars
ζ UMa (visual double, magnitudes 2.4 and 4.0, separation 14.5″)
ζ UMa and 80 UMa (Mizar and Alcor; A naked-eye double, separation 12″)

Ursa Minor (UMi, Ursae Minoris)

Meaning and origin: Little Bear. From Greek mythology – the son of Callisto (see commentary on Ursa Major). The constellation was first recorded in Ptolemy's *Almagest* in about AD 145.

This constellation contains the celestial North Pole, which has the bright star, α UMi (Polaris), located about three quarters of a degree away from it. This star therefore forms a useful marker for the approximate position of the pole in the sky.

Located on figures: 2.13, 2.22
Area: 255 square degrees
Number of stars visible to the naked eye under good conditions – 15

Named stars
Alruccabah, Cynosura or Polaris – α UMi (the brightest star, magnitude 2.0)
Kochab – β UMi
Pherkad – γ UMi
Yildun – δ UMi

Vela (Vel, Velorum)

Meaning and origin: Sails. Originally this constellation, along with Carina (keel) and Puppis (poop or stern), formed the constellation Argo (ship – also called Agro Navis and one of Ptolemy's constellations). Argo, in Greek mythology, was Jason's ship as used by the argonauts during the hunt for the golden fleece. In Lacaille's star catalogue *Coelum Australe Stelliferum* of 1763, Agro was subdivided into the three smaller constellations for convenience.

Located on figures: 2.76, 2.100, 2.117
Area: 500 square degrees
Number of stars visible to the naked eye under good conditions – 70
Brightest star – γ¹ Vel – magnitude 1.82 (archetype of the Wolf–Rayet emission line stars; a visual double star with γ² Vel which has a magnitude of 4.2 and a separation of 41″)

Named stars
Markab – κ Vel (same name as α Peg)
Al Suhail – λ Vel

Double stars
δ Vel (magnitudes 2.1 and 5.0, separation 2.5″)
γ¹ Vel (see above)
ψ Vel (visual binary, magnitudes 4.1 and 4.4, separation 1″, period 34 years)

Virgo (Vir, Virginis) – a zodiacal constellation

Meaning and origin: Virgin. From Greek mythology – either represents Demeter (Ceres), goddess of harvest, or Astraea, goddess of order and justice. The constellation was recorded in Ptolemy's *Almagest* in about AD 145, but probably pre-dates that – possibly back to Babylonian times (2000 BC).

Located on figures: 2.13, 2.43, 2.48, 2.55, 2.76, 2.122
Area: 1295 square degrees – the second largest constellation.
Number of stars visible to the naked eye under good conditions – 55

Vela – Nebulae, galaxies and other objects of interest

Object	Name	NGC	Type	RA₂₀₀₀ H	m	Dec₂₀₀₀ degrees	Visual mag	Size (′)
—	—	2547	Open Cluster	08	11	−49.3	4.7	20
Caldwell 85	—	IC 2391	Open Cluster	08	40	−53.1	2.5	50
	—	IC 2395	Open Cluster	08	41	−46.3	4.6	8
Caldwell 74	Eight-Burst	3132	Planetary Nebula	10	08	−40.4	8.2	0.8
Caldwell 79	—	3201	Globular Cluster	10	18	−46.4	6.8	18

Named stars

Spica – α Vir (the brightest star, magnitude 0.96)

Zavijava – β Vir

<u>Porrima</u> or Arich – γ Vir (visual binary, magnitudes 3.6 and 3.6, separation ranges from 0.2″ to 7″, period 170 years)

Vindemiatrix – ε Vir

Zaniah – η Vir

Syrma – ι Vir

Volans (Vol, Volantis) – originally Pisces Volans

Meaning and origin: Flying Fish. Introduced as a new constellation by Keyser and de Houtman around 1600.

Located on figures: 2.88, <u>2.100</u>, 2.117

Area: 140 square degrees

Number of stars visible to the naked eye under good conditions – 12

Brightest star – γ Vol – magnitude 3.70

Double stars

γ Vol (visual double, magnitudes 3.9 and 5.8, separation 14″)

Vulpecula (Vul, Vulpeculae) – originally Vulpecula et Anser (Fox and Goose)

Meaning and origin: Fox. Introduced as a new constellation by Hevelius in his star atlas, *Firmamentum Sobiescianum* of 1690.

Located on figures: <u>2.37</u>, <u>2.43</u>

Area: 270 square degrees

Number of stars visible to the naked eye under good conditions – 25

Brightest star – α Vul – magnitude 4.45

Virgo – Nebulae, galaxies and other objects of interest								
Object	Name	NGC	Type	RA$_{2000}$ H	m	Dec$_{2000}$ degrees	Visual mag	Size (′)
Messier 61	—	4303	Galaxy (S)	12	22	4.5	9.7	6
Messier 84	—	4374	Galaxy (E)	12	25	12.9	9.3	5
Messier 86	—	4406	Galaxy (E)	12	26	13.0	9.2	7
Messier 49	—	4472	Galaxy (E)	12	30	8.0	8.4	9
Messier 87	Virgo A	4486	Galaxy (E)	12	31	12.4	8.6	7
Messier 89	—	4552	Galaxy (E)	12	36	12.6	9.8	4
Messier 90	—	4569	Galaxy (S)	12	37	13.2	9.5	10
Messier 58	—	4579	Galaxy (S)	12	38	11.8	9.8	5
Messier 104	Sombrero	4594	Galaxy (S)	12	40	−11.6	8.3	9
Messier 59	—	4621	Galaxy (E)	12	42	11.7	9.8	5
—	—	4636	Galaxy (E)	12	43	2.7	9.6	6
Messier 60	—	4649	Galaxy (E)	12	44	11.6	8.8	7
Caldwell 52	—	4697	Galaxy (E)	12	49	−5.8	9.3	6
—	—	4699	Galaxy (S)	12	49	−8.7	9.6	4

Vulpecula – Nebulae, galaxies and other objects of interest								
Object	Name	NGC	Type	RA$_{2000}$ H	m	Dec$_{2000}$ degrees	Visual mag	Size (′)
—	Brocchi	—	Open Cluster	19	25	20.2	3.6	60
Messier 27	Dumbbell	6853	Planetary Nebula	20	00	22.7	7.6	15
Caldwell 37	—	6885	Open Cluster	20	12	26.5	5.7	7

Chapter 4

The Sky throughout the Year

The constellations visible in the sky for a particular observer depend upon the time of night, the time of year and the latitude of the observing site. Unless you are on the equator, then some of the constellations will be permanently below the horizon, and so never be seen, while others may never set (known as circumpolar constellations), and so be visible on any clear night.

There are now numerous computer programs that will plot the view from a specific site given the date and the time of night. If you have access to one of these programs then they will quickly tell you which constellations you should be able to see. The programs are advertised extensively in popular astronomy magazines (Appendix 3) and elsewhere. The astronomy magazines themselves also usually contain sky maps showing the appearance of the heavens for the following month, and similar sketches are often to be found in newspapers. Additionally there are books published annually (Appendix 3, Section A3.1) giving similar information.

However, whether or not a constellation should be visible can be determined without any of these aids, just by using the diagrams that follow. These diagrams show the rising and setting times for the major constellations, and also the times that the constellation is on the prime meridian. The prime meridian is the line passing through the zenith and the north and south points on the horizon. When a constellation is on the prime meridian it is nearest to being overhead, and so is best placed in the sky for viewing. The diagrams show the night time hours, and the times of sunrise and sunset for northern ("N"), equatorial ("Eq") and southern ("S") observers. The diagrams appear complex to start with, but they are actually easy to

use, and with a little practice you will soon become familiar with them.

To find whether a particular constellation is visible or not and roughly whereabouts it may be found in the sky, then go to the diagram for that constellation (Table 4.1). Rule a vertical line at the point corresponding to the date. If the vertical line intersects any of the lines marked as "Rising Times", then the time at which the constellation rises may be read off the vertical axis. On most diagrams three rising lines are given corresponding to latitudes 40°N (marked "N"), the equator (marked "Eq") and 40°S (marked "S"). Similarly if the vertical line intersects any of the lines marked "Setting Times", then the time at which the constellation sets may be found. If the vertical line intersects the line marked "Prime Meridian", then that gives the time at which the constellation will be highest in the sky. The times for the minor constellations may be found from those for their nearest major constellation.

Thus, for example, in Figure 4.1 Auriga is shown for October 15th. For northern hemisphere observers, the constellation will then rise at about 19.00 hours (7 pm), for equatorial observers it will rise at about 22.00 hours (10 pm), and for southern hemisphere observers it will rise at about 01.00 hours (1 am). For all observers it will be on the prime meridian at about 04.00 hours (4 am). For no observer will it set before the Sun has risen. On June 1st, however, Auriga will have set before sunset for southern observers, but will still be briefly visible to equatorial and northern observers; setting at about 19.00 hours (7 pm), and 22.00 hours (10 pm) respectively. For southern observers, Auriga will not be visible at all between about May 15th (when it sets with the Sun), and

Table 4.1. The major constellations and their visibility diagrams

Constellation	Diagram	Constellation	Diagram
Andromeda	Fig. 4.2	Aquila	Fig. 4.3
Ara	Fig. 4.4	Auriga	Fig. 4.5
Boötes	Fig. 4.6	Canis Major	Fig. 4.7
Canis Minor	Fig. 4.8	Carina	Fig. 4.9
Cassiopeia	Fig. 4.10	Centaurus	Fig. 4.11
Cetus	Fig. 4.12	Crux	Fig. 4.13
Cygnus	Fig. 4.14	Gemini	Fig. 4.15
Grus	Fig. 4.16	Leo	Fig. 4.17
Lupus	Fig. 4.18	Lyra	Fig. 4.19
Orion	Fig. 4.20	Pavo	Fig. 4.21
Pegasus	Fig. 4.22	Perseus	Fig. 4.23
Pisces Austrinus	Fig. 4.24	Puppis	Fig. 4.25
Sagittarius	Fig. 4.26	Scorpius	Fig. 4.27
Taurus	Fig. 4.28	Triangulum Australe	Fig. 4.29
Ursa Major	Fig. 4.30	Ursa Minor	Fig. 4.31
Vela	Fig. 4.32		

July 10th (when it rises with the Sun). Whereas for northern hemisphere observers it will always be visible for some part of the night, and will be visible all night between about November 10th and January 8th.

The times plotted in Figs 4.2 to 4.32 are based upon the centres of the constellations, and are normal civil times (that is, not including any adjustments for summer time etc.). Since some constellations span tens of degrees, parts of the constellation may be visible earlier than the rising times given by the diagrams, or later than the setting times. The times are plotted only for latitudes 0° and ±40°. If your latitude differs from any of these values, then you will need to interpolate or extrapolate from the lines shown on the diagrams. However, the actual times of rising or setting may differ from those shown by up to 30 minutes if you are towards the eastern or western edge of your local time zone, so the times should only be regarded as an approximate guide.

To get an idea of the constellation's position in the sky, refer to the line marked "Prime Meridian". If you are observing before the time given by that line, then the constellation will be towards the east, while after that time it

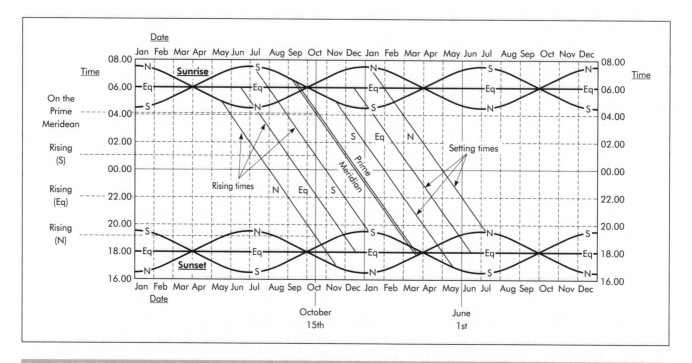

Figure 4.1. Visibility diagram for Auriga (circumpolar for latitudes north of about +50°, not seen south of latitude about –50°). Examples are shown for October 15th and June 1st.

will be towards the west. On some maps only one rising or setting line is marked, corresponding to the time for equatorial observers. This occurs when the constellation is circumpolar at 40° N or 40° S. It will then be visible all night and at all times of the year. However for observers above or below the corresponding latitude in the opposite hemisphere, the constellation will not be seen at all.

The best time to observe any constellation is when it is highest in the sky; that is, when it is on or near the prime meridian. Since most people will probably want to observe in the evening, different constellations are best placed for observing at different times of the year. The months when the major constellations are on or near the prime meridian at 22.00 hours (10 pm) are listed in Table 4.2.

Table 4.2. Times of the year when constellations are on or near the prime meridian at 22.00 hours (10 pm)

January	Taurus, Auriga, Orion
February	Canis Major, Gemini, Canis, Minor, Puppis
March	Carina, Vela
April	Leo, Ursa Major
May	Crux, Centaurus, Sagittarius
June	Boötes, Ursa Minor, Lupus, Triangulum Australe
July	Scorpius, Ara
August	Lyra, Aquila, Pavo
September	Cygnus
October	Grus, Pisces Austrinus, Pegasus
November	Andromeda, Cassiopeia, Cetus
December	Perseus

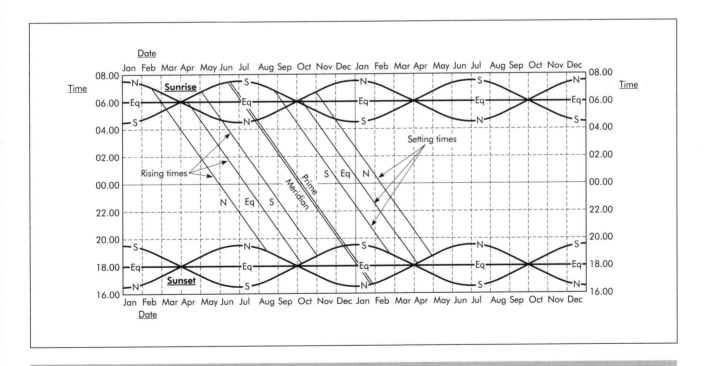

Figure 4.2. Visibility diagram for Andromeda (circumpolar for latitudes north of about +55°, not seen south of latitudes about −55°).

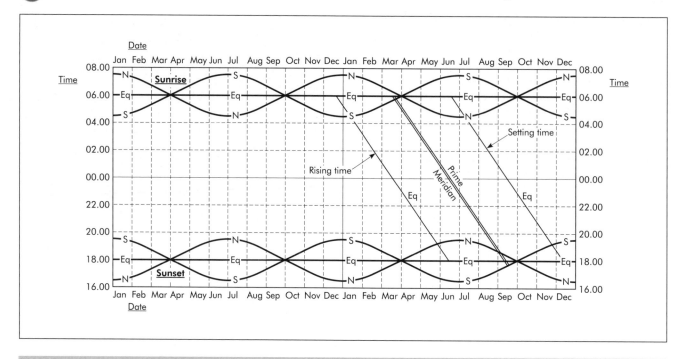

Figure 4.3. Visibility diagram for Ara (circumpolar for latitudes south of about –35°, not seen north of latitudes about +35°).

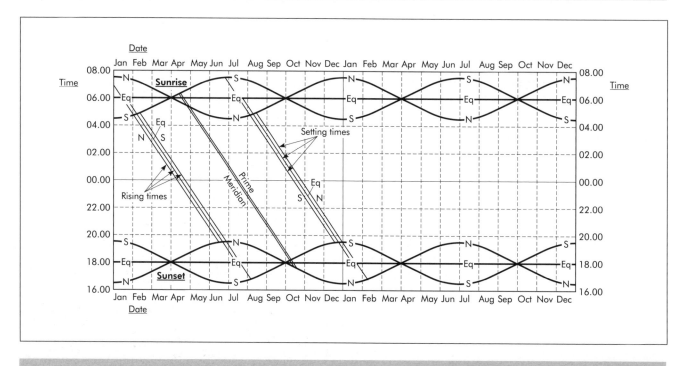

Figure 4.4. Visibility diagram for Aquila (an equatorial constellation).

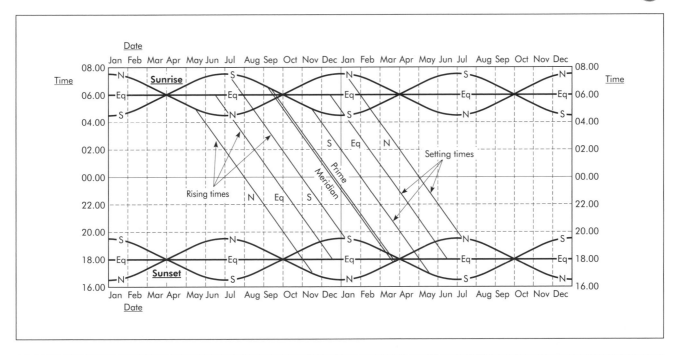

Figure 4.5. Visibility diagram for Auriga (circumpolar for latitudes north of about +50°, not seen south of latitudes about –50°).

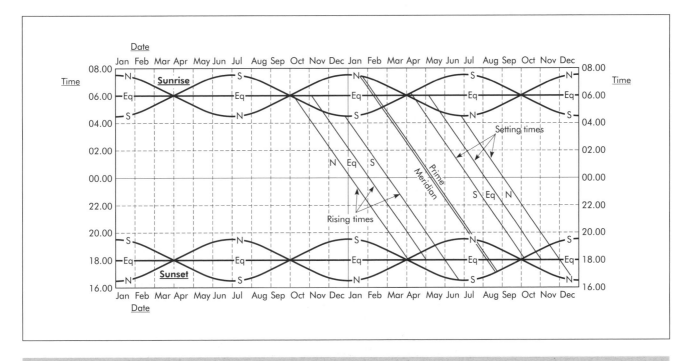

Figure 4.6. Visibility diagram for Boötes (circumpolar for latitudes north of about +60°, not seen south of latitudes about –60°).

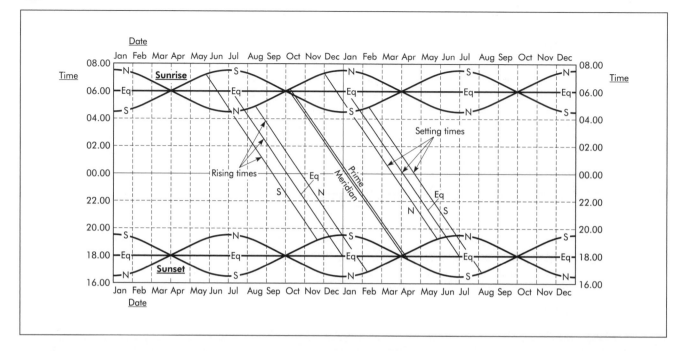

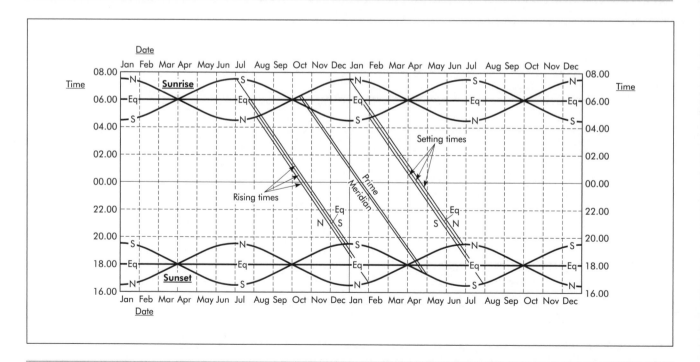

Figure 4.8. Visibility diagram for Canis Minor (an equatorial constellation).

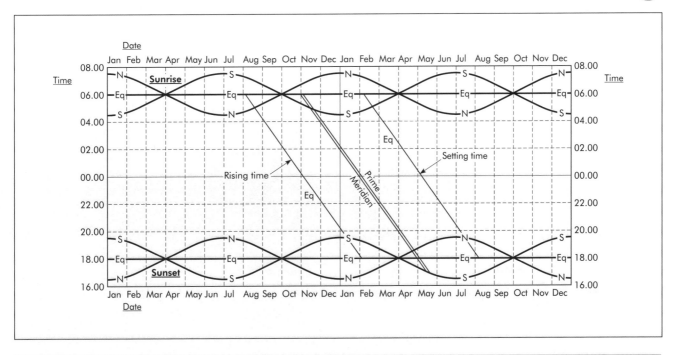

Figure 4.9. Visibility diagram for Carina (circumpolar for latitudes south of about –30°, not seen north of latitudes about +30°).

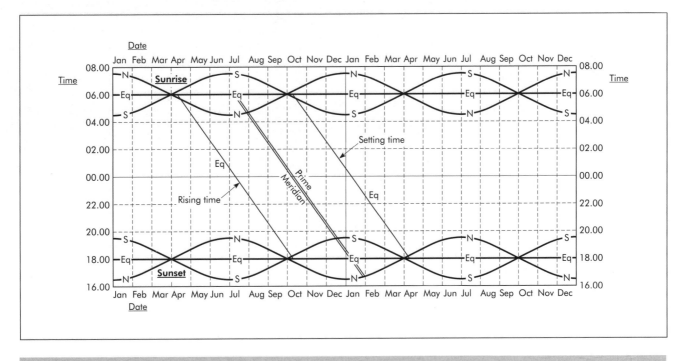

Figure 4.10. Visibility diagram for Cassiopeia (circumpolar for latitudes north of about +30°, not seen south of latitudes about –30°).

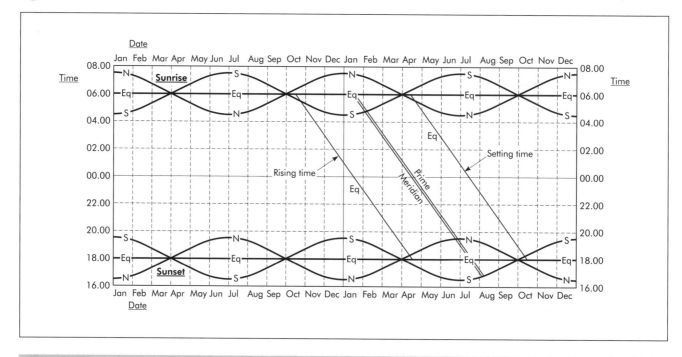

Figure 4.11. Visibility diagram for Centaurus (circumpolar for latitudes south of about –35°, not seen north of latitudes about +35°).

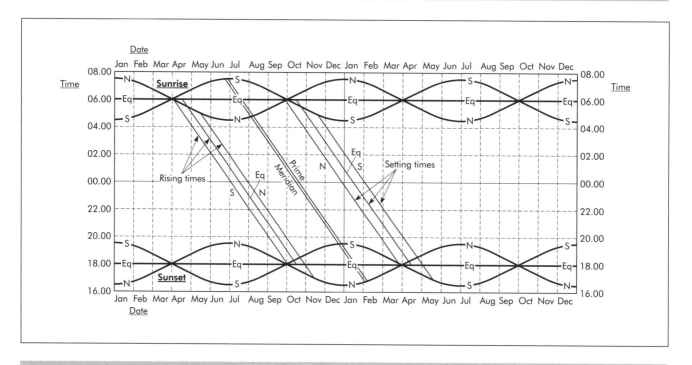

Figure 4.12. Visibility diagram for Cetus (an equatorial constellation).

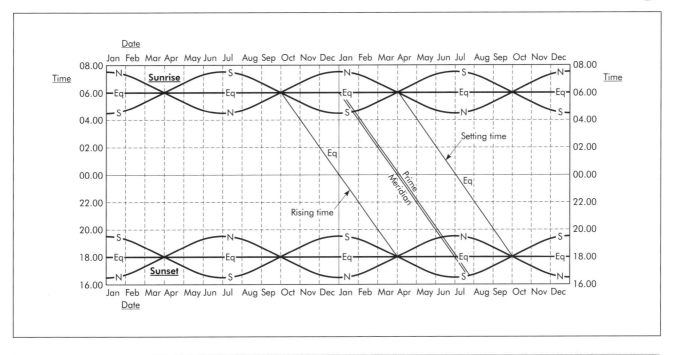

Figure 4.13. Visibility diagram for Crux (circumpolar for latitudes south of about –30°, not seen north of latitudes about +30°).

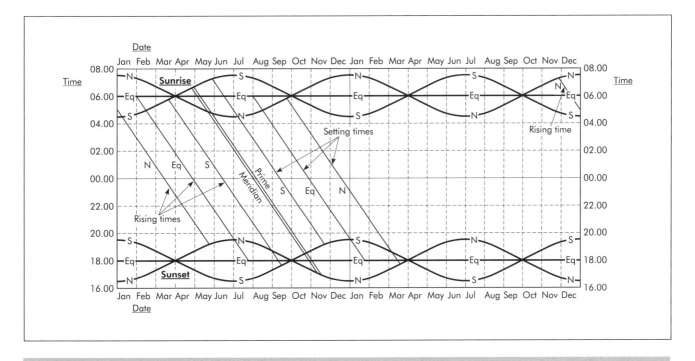

Figure 4.14. Visibility diagram for Cygnus (circumpolar for latitudes north of about +50°, not seen south of latitudes about –50°).

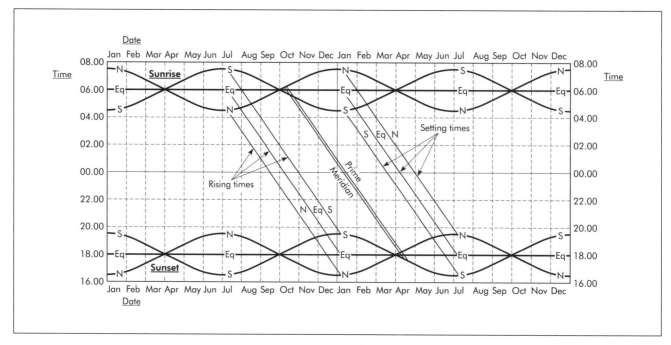

Figure 4.15. Visibility diagram for Gemini (circumpolar for latitudes north of about +65°, not seen south of latitudes about −65°).

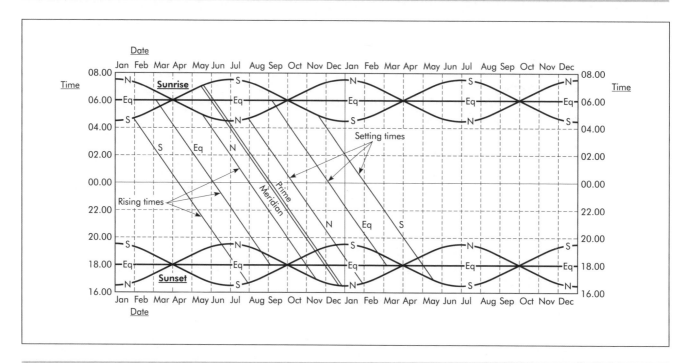

Figure 4.16. Visibility diagram for Grus (circumpolar for latitudes south of about −45°, not seen north of latitudes about +45°).

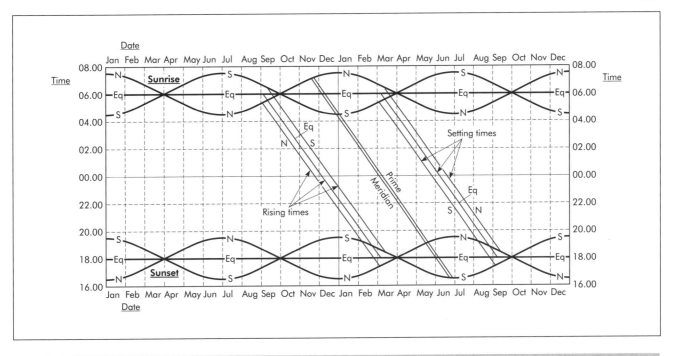

Figure 4.17. Visibility diagram for Leo (an equatorial constellation).

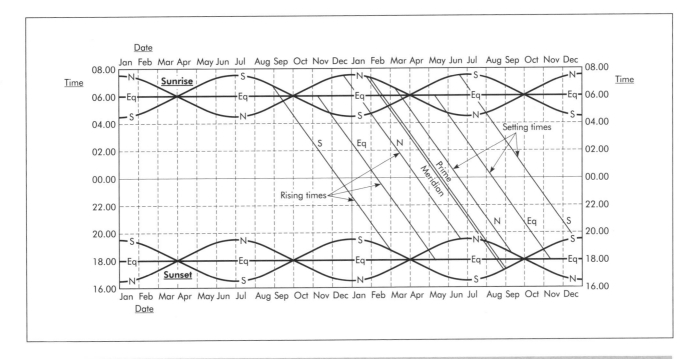

Figure 4.18. Visibility diagram for Lupus (circumpolar for latitudes south of about –45°, not seen north of latitudes about +45°).

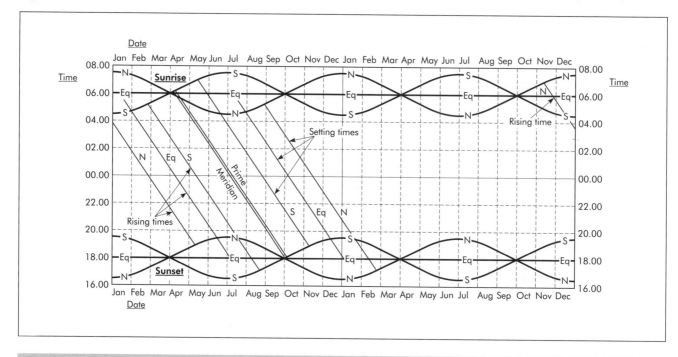

Figure 4.19. Visibility diagram for Lyra (circumpolar for latitudes north of about +55°, not seen south of latitudes about –55°).

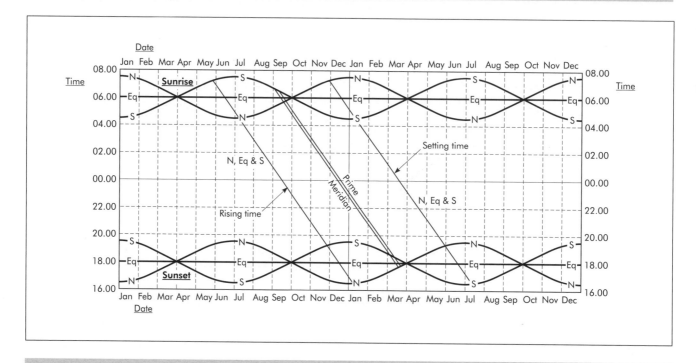

Figure 4.20. Visibility diagram for Orion (an equatorial constellation).

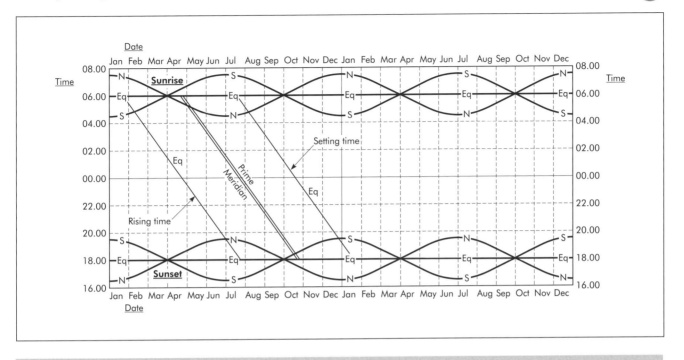

Figure 4.21. Visibility diagram for Pavo (circumpolar for latitudes south of about –25°, not seen north of latitudes about +25°).

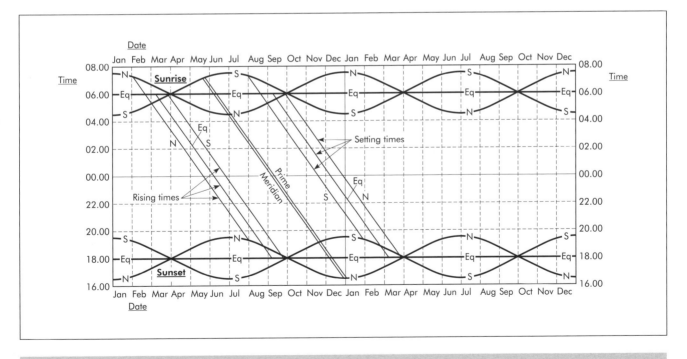

Figure 4.22. Visibility diagram for Pegasus (circumpolar for latitudes north of about +70°, not seen south of latitudes about –70°).

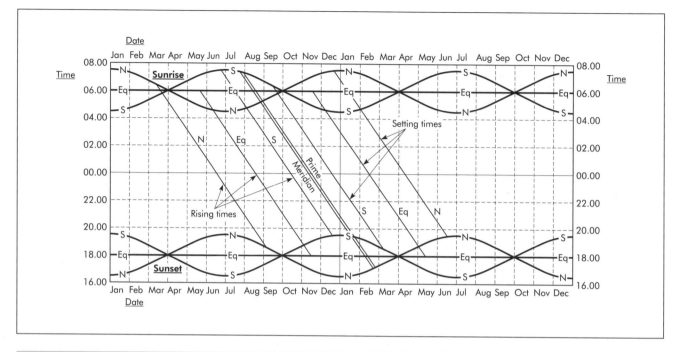

Figure 4.23. Visibility diagram for Perseus (circumpolar for latitudes north of about +45°, not seen south of latitudes about −45°).

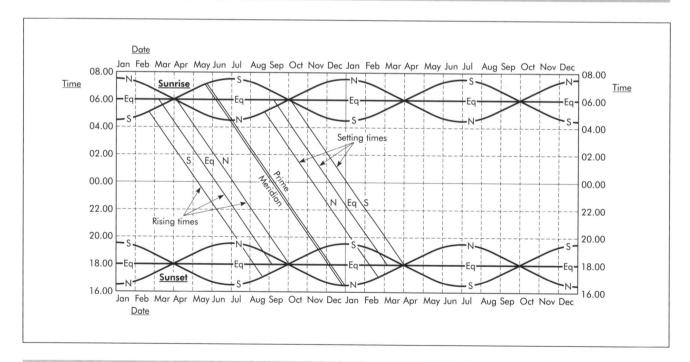

Figure 4.24. Visibility diagram for Pisces Austrinus (circumpolar for latitudes south of about −60°, not seen north of latitudes about +60°).

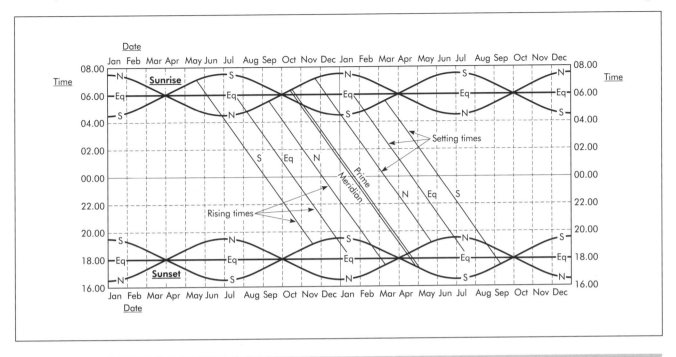

Figure 4.25. Visibility diagram for Puppis (circumpolar for latitudes south of about −50°, not seen north of latitudes about +50°).

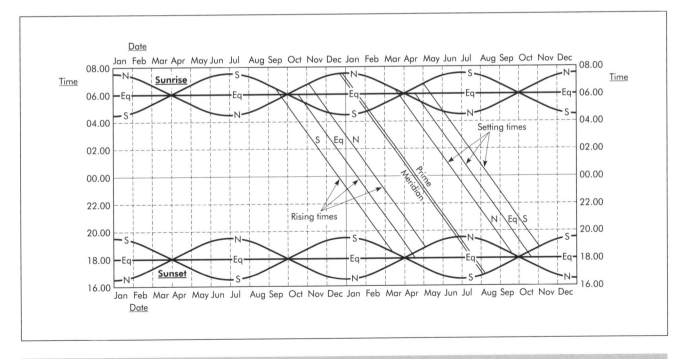

Figure 4.26. Visibility diagram for Sagittarius (circumpolar for latitudes south of about −60°, not seen north of latitudes about +60°).

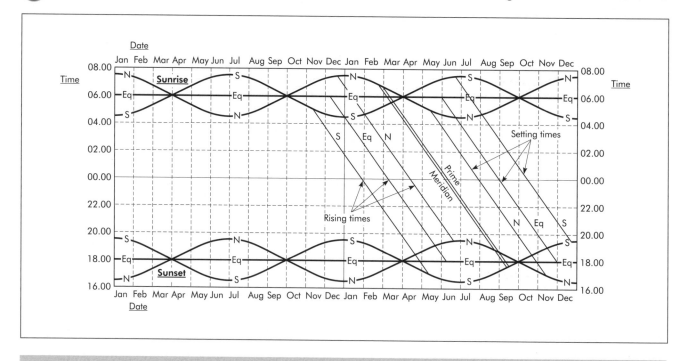

Figure 4.27. Visibility diagram for Scorpius (circumpolar for latitudes south of about –55°, not seen north of latitudes about +55°).

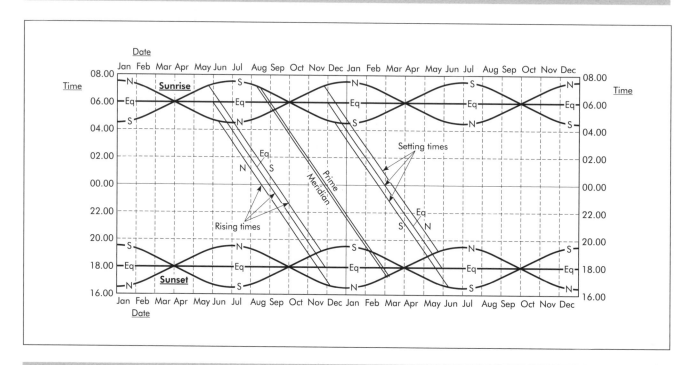

Figure 4.28. Visibility diagram for Taurus (an equatorial constellation).

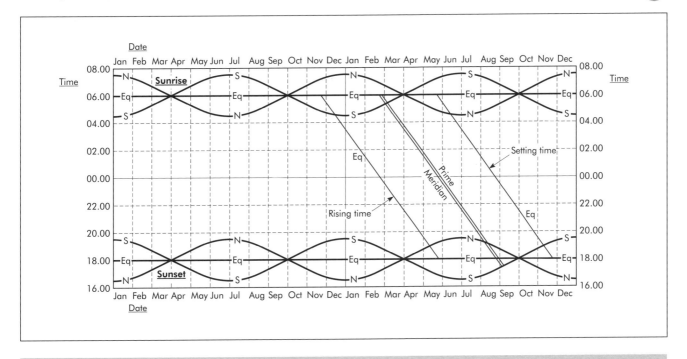

Figure 4.29. Visibility diagram for Triangulum Australe (circumpolar for latitudes south of about −25°, not seen north of latitudes about +25°).

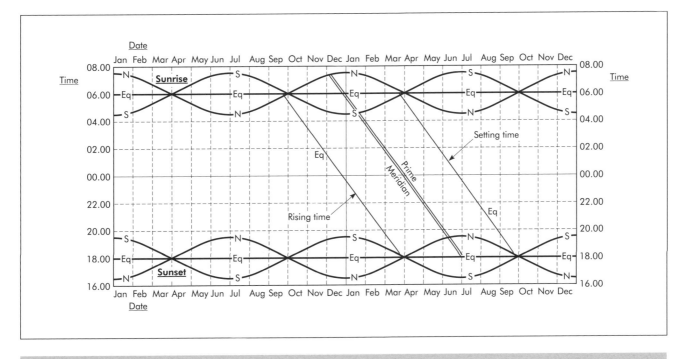

Figure 4.30. Visibility diagram for Ursa Major (circumpolar for latitudes north of about +35°, not seen south of latitudes about −35°).

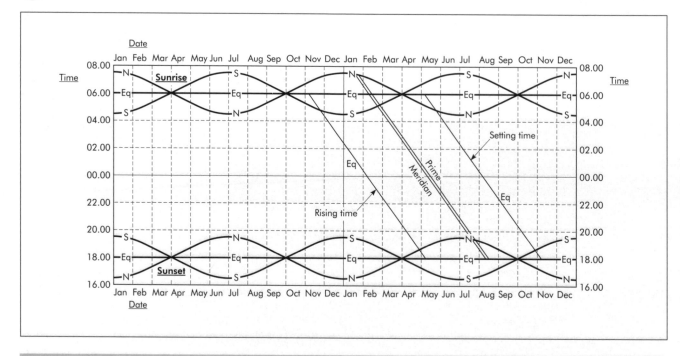

Figure 4.31. Visibility diagram for Ursa Minor (circumpolar for latitudes north of about +10°, not seen south of latitudes about –10°).

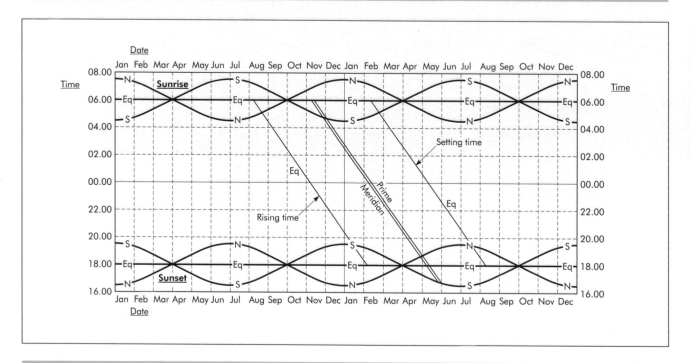

Figure 4.32. Visibility diagram for Vela (circumpolar for latitudes south of about –40°, not seen north of latitudes about +40°).

The Messier and Caldwell Objects

Charles Messier (1730–1817) was an observational astronomer working from Paris in the eighteenth century. He discovered between 15 and 21 comets and observed many more. During his observations he encountered nebulous objects that were not comets. Some of these objects were his own discoveries, while others had been known before. In 1774 he published a list of 45 of these nebulous objects. His purpose in publishing the list was so that other comet-hunters should not confuse the nebulae with comets. Over the following decades he published supplements which increased the number of objects in his catalogue to 103 though objects M101 and M102 were in fact the same. Later other astronomers added a replacement for M102 and objects 104 to 110. It is now thought probable that Messier had observed these later additions with the exception of the last. Thus the modern version of his catalogue has 109 objects in it. Several of the objects in Messier's original lists are now difficult to identify, and "best guesses" have had to be made regarding the objects intended. Ironically in one case (M91) it is possible that Messier's original observation was of an unrecognised comet!

Messier observed mostly with 3 to 3.5 inch (75–90 mm) refractors. He had access to a 7.5 inch (190 mm) Gregorian reflector but since this used mirrors made from speculum metal, its equivalent aperture would have been only 3 inches or so. With modern telescopes it should be possible to observe all the Messier objects with a 2.5 inch (60 mm) or larger instrument. This accessibility of the objects in Messier's list, compared with Herschel's *New General Catalogue* (NGC) which was being compiled at the same time as Messier's observations but using much larger telescopes, probably explains its modern popularity. It is a challenging but achievable task for most amateur astronomers to observe all the Messier objects. At "star parties" and within astronomy clubs, going for the maximum number of Messier objects observed is a popular competition. Indeed at some times of the year it is just about possible to observe most of them in a single night.

Messier observed from Paris and therefore the most southerly object in his list is M7 in Scorpius with a declination of –35°. He also missed several objects from his list such as h and χ Per and the Hyades which most observers would feel should have been included. The well-known astronomer Dr Patrick Moore has therefore recently introduced the *Caldwell Catalogue* (his full name is Patrick *Caldwell*-Moore). This has 109 objects like Messier's list, but covers the whole sky. The Caldwell objects are listed in decreasing order of declination, so that from a given latitude all objects from C1 to C*n* (or from C109 down to C*n* for southern observers), where *n* is the number of the most southerly (northerly) objects rising at that site, should be visible. There is no overlap between the Messier and Caldwell Catalogues, and the two taken together (see below) will give sufficient fascinating and spectacular objects to keep most astronomers occupied for several years' worth of observing.

The Messier and Caldwell objects are listed within each constellation as it is discussed in Section 3.2. Here they are listed in order for ease of reference.

Table A1.1. The Messier Objects

Object	Name	NGC	Type	RA₂₀₀₀ H	m	Dec₂₀₀₀ degrees	Constellation	Visual* mag	Size** (')
Messier 1	Crab	1952	Supernova Remnant	05	35	22.0	Tau	8.4	6
Messier 2		7089	Globular Cluster	21	34	−0.8	Aqr	6.5	13
Messier 3		5272	Globular Cluster	13	42	28.4	CVn	6.4	16
Messier 4		6121	Globular Cluster	16	24	−26.5	Sco	5.9	26
Messier 5		5904	Globular Cluster	15	19	2.1	Ser Cap	5.8	17
Messier 6	Butterfly	6405	Open Cluster	17	40	−32.2	Sco	4.2	15
Messier 7		6475	Open Cluster	17	54	−34.8	Sco	3.3	80
Messier 8	Lagoon	6523	Emission Nebula	18	04	−24.4	Sgr	5.8	90
Messier 9		6333	Globular Cluster	17	19	−18.5	Oph	7.9	9
Messier 10		6254	Globular Cluster	16	57	−4.1	Oph	6.6	15
Messier 11	Wild Duck	6705	Open Cluster	18	51	−6.3	Sct	5.8	14
Messier 12		6218	Globular Cluster	16	47	−2.0	Oph	6.6	15
Messier 13		6205	Globular Cluster	16	42	36.5	Her	5.9	17
Messier 14		6402	Globular Cluster	17	38	−3.3	Oph	7.6	12
Messier 15		7078	Globular Cluster	21	30	12.2	Peg	6.4	12
Messier 16	Eagle	6611	Open Cluster	18	19	−13.8	Ser	6.0	7
Messier 17	Omega	6618	Emission Nebula	18	21	−16.2	Sgr	6.0	46
Messier 18		6613	Open Cluster	18	20	−17.1	Sgr	6.9	9
Messier 19		6273	Globular Cluster	17	03	−26.3	Oph	7.2	14
Messier 20	Trifid	6514	Emission Nebula	18	03	−23.0	Sgr	6.3	29
Messier 21		6531	Open Cluster	18	05	−22.5	Sgr	5.9	13
Messier 22		6656	Globular Cluster	18	36	−23.9	Sgr	5.1	24
Messier 23		6494	Open Cluster	17	57	−19.0	Sgr	5.5	27
Messier 24		6603	Star Cloud	18	18	−18.4	Sgr	11.1	5
Messier 25		IC 4725	Open Cluster	18	32	−19.3	Sgr	4.6	32
Messier 26		6694	Open Cluster	18	45	−9.4	Sct	8.0	15
Messier 27	Dumbbell	6853	Planetary Nebula	20	00	22.7	Vul	7.6	15
Messier 28		6626	Globular Cluster	18	25	−24.9	Sgr	6.9	11
Messier 29		6913	Open Cluster	20	24	38.5	Cyg	6.6	7
Messier 30		7099	Globular Cluster	21	40	−23.2	Cap	7.5	11
Messier 31	Andromeda	224	Galaxy	00	43	41.3	And	3.5	180
Messier 32		221	Galaxy	00	43	40.9	And	8.2	8
Messier 33	Triangulum	598	Galaxy	01	34	30.7	Tri	5.7	62
Messier 34		1039	Open Cluster	02	42	42.8	Per	5.2	35
Messier 35		2168	Open Cluster	06	09	24.3	Gem	5.1	28
Messier 36		1960	Open Cluster	05	36	34.1	Aur	6.0	12
Messier 37		2099	Open Cluster	05	52	32.6	Aur	5.6	24
Messier 38		1912	Open Cluster	05	29	35.8	Aur	6.4	21
Messier 39		7092	Open Cluster	21	32	48.4	Cyg	4.6	32
Messier 40			2 stars	12	22	58.1	UMa	0.8	9
Messier 41		2287	Open Cluster	06	47	−20.7	CMa	4.5	38
Messier 42	Orion	1976	Emission Nebula	05	35	−5.5	Ori	4.0	66
Messier 43	Orion	1982	Emission Nebula	05	36	−5.3	Ori	9.0	20
Messier 44	Praesepe	2632	Open Cluster	08	40	20.0	Cnc	3.1	95
Messier 45	Pleiades		Open Cluster	03	47	24.1	Tau	1.2	110
Messier 46		2437	Open Cluster	07	42	−14.8	Pup	6.1	27
Messier 47		2422	Open Cluster	07	37	−14.5	Pup	4.4	30
Messier 48		2548	Open Cluster	08	14	−5.8	Hya	5.8	54
Messier 49		4472	Galaxy	12	30	8.0	Vir	8.4	9
Messier 50		2323	Open Cluster	07	03	−8.3	Mon	5.9	16

Table A1.1. (continued)

Object	Name	NGC	Type	RA₂₀₀₀ H	m	Dec₂₀₀₀ degrees	Constellation	Visual* mag	Size** (')
Messier 51	Whirlpool	5194	Galaxy	13	30	47.2	CVn	8.4	11
Messier 52		7654	Open Cluster	23	24	61.6	Cas	6.9	13
Messier 53		5024	Globular Cluster	13	13	18.2	Com	7.7	13
Messier 54		6715	Globular Cluster	18	55	−30.5	Sgr	7.7	9
Messier 55		6809	Globular Cluster	19	40	−31.0	Sgr	7.0	19
Messier 56		6779	Globular Cluster	19	17	30.2	Lyr	8.3	7
Messier 57	Ring	6720	Planetary Nebula	18	54	33.0	Lyr	9.7	3
Messier 58		4579	Galaxy	12	38	11.8	Vir	9.8	5
Messier 59		4621	Galaxy	12	42	11.7	Vir	9.8	5
Messier 60		4649	Galaxy	12	44	11.6	Vir	8.8	7
Messier 61		4303	Galaxy	12	22	4.5	Vir	9.7	6
Messier 62		6266	Globular Cluster	17	01	−30.1	Oph	6.6	14
Messier 63	Sunflower	5055	Galaxy	13	16	42.0	CVn	8.6	12
Messier 64	Black-eye	4826	Galaxy	12	57	21.7	Com	8.5	9
Messier 65		3623	Galaxy	11	19	13.1	Leo	9.3	10
Messier 66		3627	Galaxy	11	20	13.0	Leo	9.0	9
Messier 67		2682	Open Cluster	08	50	11.8	Cnc	6.9	30
Messier 68		4590	Globular Cluster	12	40	−26.8	Hya	8.2	12
Messier 69		6637	Globular Cluster	18	31	−32.4	Sgr	7.7	7
Messier 70		6681	Globular Cluster	18	43	−32.3	Sgr	8.8	8
Messier 71		6838	Globular Cluster	19	54	18.8	Sge	8.3	7
Messier 72		6981	Globular Cluster	20	54	−12.5	Aqr	9.4	6
Messier 73		6994	Open Cluster	20	59	−12.6	Aqr	8.9	3
Messier 74		628	Galaxy	01	37	15.8	Psc	9.2	10
Messier 75		6864	Globular Cluster	20	06	−21.9	Sgr	8.6	6
Messier 76	Little Dumbbell	650	Planetary Nebula	01	42	51.6	Per	12.2	5
Messier 77		1068	Galaxy	02	43	0.0	Cet	8.8	7
Messier 78		2068	Reflection Nebula	05	47	0.1	Ori	8.0	8
Messier 79		1904	Globular Cluster	05	25	−24.6	Lep	8.0	9
Messier 80		6093	Globular Cluster	16	17	−23.0	Sco	7.2	9
Messier 81	Bode's	3031	Galaxy	09	56	69.1	UMa	6.9	26
Messier 82		3034	Galaxy	09	56	69.7	UMa	8.4	11
Messier 83		5236	Galaxy	13	37	−29.9	Hya	8.2	11
Messier 84		4374	Galaxy	12	25	12.9	Vir	9.3	5
Messier 85		4382	Galaxy	12	25	18.2	Com	9.2	7
Messier 86		4406	Galaxy	12	26	13.0	Vir	9.2	7
Messier 87	Virgo A	4486	Galaxy	12	31	12.4	Vir	8.6	7
Messier 88		4501	Galaxy	12	32	14.4	Com	9.5	7
Messier 89		4552	Galaxy	12	36	12.6	Vir	9.8	4
Messier 90		4569	Galaxy	12	37	13.2	Vir	9.5	10
Messier 91		4548	Galaxy	12	35	14.5	Com	10.2	5
Messier 92		6341	Globular Cluster	17	17	43.1	Her	6.5	11
Messier 93		2447	Open Cluster	07	45	−23.9	Pup	6.2	22
Messier 94		4736	Galaxy	12	51	41.1	CVn	8.2	11
Messier 95		3351	Galaxy	10	44	11.7	Leo	9.7	7
Messier 96		3368	Galaxy	10	47	11.8	Leo	9.2	7
Messier 97	Owl	3587	Planetary Nebula	11	15	55.0	UMa	12	3
Messier 98		4192	Galaxy	12	14	14.9	Com	10.1	10
Messier 99		4254	Galaxy	12	19	14.4	Com	9.8	5
Messier 100		4321	Galaxy	12	23	15.8	Com	9.4	7

Table A1.1. (continued)

Object	Name	NGC	Type	RA2000 H	m	Dec2000 degrees	Constellation	Visual* mag	Size** (')
Messier 101	Pinwheel	5457	Galaxy	14	03	54.4	UMa	7.7	27
Messier 102		5866	Galaxy	15	07	55.8	Dra	10.0	5
Messier 103		581	Open Cluster	01	33	60.7	Cas	7.4	6
Messier 104	Sombrero	4594	Galaxy	12	40	−11.6	Vir	8.3	9
Messier 105		3379	Galaxy	10	48	12.6	Leo	9.3	5
Messier 106		4258	Galaxy	12	19	47.3	CVn	8.3	18
Messier 107		6171	Globular Cluster	16	33	−13.1	Oph	8.1	10
Messier 108		3556	Galaxy	11	12	55.7	UMa	10.1	8
Messier 109		3992	Galaxy	11	58	53.4	UMa	9.8	8

* This is the integrated magnitude over the whole area of the object. An angularly large object with a bright magnitude may therefore be less easy to see than a smaller object with a fainter magnitude. The magnitudes of the emission nebulae in particular may be misleading because they frequently contain brighter and darker regions.

** This is the largest dimension of the object. Some objects may be filamentary or have a brighter core or outer region making them easier to see than might be expected.

(Data for this table obtained from *Sky Catalogue 2000.0, Vol. 2* (Ed. A. Hirshfeld and R.W. Sinnott, Cambridge University Press, 1985); *Astrophysical Quantities* (C.W. Allen, Athlone Press, 1973); *NGC 2000.0* (R.W. Sinnott, Cambridge University Press, 1988); *Visual Astronomy of the Deep Sky* (R.N. Clark, Cambridge University Press, 1990); *Hartung's Astronomical Objects for Southern Telescopes* (D. Malin and D.J. Frew, Cambridge University Press, 1995); *Astrophysical Journal Supplement*, **4,** 257, 1959, S. Sharpless.)

Table A1.2. The Caldwell Objects

Object	Name	NGC	Type	RA$_{2000}$ H	m	Dec$_{2000}$ degrees	Constellation	Visual* mag	Size** (')
Caldwell 1		188	Open Cluster	00	44	85.3	Cep	8.1	14
Caldwell 2		40	Planetary Nebula	00	13	72.5	Cep	10.7	0.6
Caldwell 3		4236	Galaxy	12	17	69.5	Dra	9.7	19
Caldwell 4		7023	Reflection Nebula	21	02	68.2	Cep	7.0	18
Caldwell 5		IC 342	Galaxy	03	47	68.1	Cam	9.1	18
Caldwell 6	Cat's Eye	6543	Planetary Nebula	17	59	66.6	Dra	8.8	6
Caldwell 7		2403	Galaxy	07	37	65.6	Cam	8.4	18
Caldwell 8		559	Open Cluster	01	30	63.3	Cas	9.5	5
Caldwell 9	Cave	Sh2-155	Emission Nebula	22	57	62.6	Cep	≈ 9	50
Caldwell 10		663	Open Cluster	01	46	61.3	Cas	7.1	16
Caldwell 11	Bubble	7635	Emission Nebula	23	21	61.2	Cas	8.5	15
Caldwell 12		6946	Galaxy	20	35	60.2	Cep	8.9	11
Caldwell 13		457	Open Cluster	01	19	58.3	Cas	6.4	13
Caldwell 14	h & χ Per	869/884	Open Cluster	02	20	57.1	Per	4.3/4.4	30/30
Caldwell 15	Blinking	6826	Planetary Nebula	19	45	50.5	Cyg	9.8	2
Caldwell 16		7243	Open Cluster	22	15	49.9	Lac	6.4	21
Caldwell 17		147	Galaxy	00	33	48.5	Cas	9.3	13
Caldwell 18		185	Galaxy	00	39	48.3	Cas	9.2	12
Caldwell 19	Cocoon	IC 5146	Emission Nebula	21	54	47.3	Cyg	7.2	12
Caldwell 20	North America	7000	Emission Nebula	20	59	44.3	Cyg	5.0	120
Caldwell 21		4449	Galaxy	12	28	44.1	CVn	9.4	5
Caldwell 22	Blue Snowball	7662	Planetary Nebula	23	26	42.6	And	9.2	2
Caldwell 23		891	Galaxy	02	23	42.4	And	10.0	14
Caldwell 24		1275	Galaxy	03	20	41.5	Per	11.6	3
Caldwell 25		2419	Globular Cluster	07	38	38.9	Lyn	10.4	4
Caldwell 26		4244	Galaxy	12	18	37.8	CVn	10.2	16
Caldwell 27	Crescent	6888	Emission Nebula	20	12	38.4	Cyg	≈ 11	20
Caldwell 28		752	Open Cluster	01	58	37.7	And	5.7	50
Caldwell 29		5005	Galaxy	13	11	37.1	CVn	9.8	5
Caldwell 30		7331	Galaxy	22	37	34.4	Peg	9.5	11
Caldwell 31	Flaming Star	IC 405	Emission Nebula	05	16	34.3	Aur	≈ 7	30
Caldwell 32		4631	Galaxy	12	42	32.5	CVn	9.3	15
Caldwell 33	Veil (E)	6992/5	Supernova Remnant	20	57	31.5	Cyg	8.0	60
Caldwell 34	Veil (W)	6960	Supernova Remnant	20	46	30.7	Cyg	8.0	70
Caldwell 35		4889	Galaxy	13	00	28.0	Com	11.4	3
Caldwell 36		4559	Galaxy	12	36	28.0	Com	9.9	11
Caldwell 37		6885	Open Cluster	20	12	26.5	Vul	5.7	7
Caldwell 38		4565	Galaxy	12	36	26.0	Com	9.6	16
Caldwell 39	Eskimo	2392	Planetary Nebula	07	29	20.9	Gem	9.9	0.7
Caldwell 40		3626	Galaxy	11	20	18.4	Leo	10.9	3
Caldwell 41	Hyades		Open Cluster	04	27	16.0	Tau	0.5	330
Caldwell 42		7006	Globular Cluster	21	02	16.2	Del	10.6	3
Caldwell 43		7814	Galaxy	00	03	16.2	Peg	10.5	6
Caldwell 44		7479	Galaxy	23	05	12.3	Peg	11.0	4
Caldwell 45		5248	Galaxy	13	38	8.9	Boö	10.2	7
Caldwell 46	Hubble's variable	2261	Emission Nebula	06	39	8.7	Mon	10.0	2
Caldwell 47		6934	Globular Cluster	20	34	7.4	Del	8.9	6
Caldwell 48		2775	Galaxy	09	10	7.0	Cnc	10.3	5
Caldwell 49	Rosette	2237–9	Emission Nebula	06	32	5.1	Mon	≈ 4	80
Caldwell 50		2244	Open Cluster	06	32	4.9	Mon	4.8	24

Table A1.2. (continued)

Object	Name	NGC	Type	RA$_{2000}$ H m	Dec$_{2000}$ degrees	Constellation	Visual* mag	Size** (')
Caldwell 51		IC 1613	Galaxy	01 05	2.1	Cet	9.3	12
Caldwell 52		4697	Galaxy	12 49	−5.8	Vir	9.3	6
Caldwell 53	Spindle	3115	Galaxy	10 05	−7.7	Sex	9.2	8
Caldwell 54		2506	Open Cluster	08 00	−10.8	Mon	7.6	7
Caldwell 55	Saturn	7009	Planetary Nebula	21 04	−11.4	Aqr	8.3	2
Caldwell 56		246	Planetary Nebula	00 47	−11.9	Cet	8.0	4
Caldwell 57	Barnard's	6822	Galaxy	19 45	−14.8	Sgr	9.4	10
Caldwell 58		2360	Open Cluster	07 18	−15.6	CMa	7.2	13
Caldwell 59	Ghost of Jupiter	3242	Planetary Nebula	10 25	−18.6	Hya	8.6	21
Caldwell 60	Antennae	4038	Galaxy	12 02	−18.9	Crv	10.7	3
Caldwell 61	Antennae	4039	Galaxy	12 02	−18.9	Crv	10.7	3
Caldwell 62		247	Galaxy	00 47	−20.8	Cet	8.9	20
Caldwell 63	Helix	7293	Planetary Nebula	22 30	−20.8	Aqr	7.4	13
Caldwell 64		2362	Open Cluster	07 19	−25.0	CMa	4.1	8
Caldwell 65	Silver Coin	253	Galaxy	00 48	−25.3	Scl	7.1	25
Caldwell 66		5694	Globular Cluster	14 40	−26.5	Hya	10.2	4
Caldwell 67		1097	Galaxy	02 46	−30.3	For	9.3	9
Caldwell 68	R CrA	6729	Reflection Nebula	19 02	−37.0	CrA	≈ 11	1
Caldwell 69	Bug	6302	Planetary Nebula	17 14	−37.1	Sco	12.8	1
Caldwell 70		300	Galaxy	00 55	−37.7	Scl	8.7	20
Caldwell 71		2477	Open Cluster	07 52	−38.6	Pup	5.8	27
Caldwell 72		55	Galaxy	00 15	−39.2	Scl	7.9	32
Caldwell 73		1851	Globular Cluster	05 14	−40.1	Col	7.3	11
Caldwell 74	Eight-Burst	3132	Planetary Nebula	10 08	−40.4	Vel	8.2	0.8
Caldwell 75		6124	Open Cluster	16 26	−40.7	Sco	5.8	29
Caldwell 76		6231	Open Cluster	16 54	−41.8	Sco	2.6	15
Caldwell 77	Cen A	5128	Galaxy	13 26	−43.0	Cen	7.0	18
Caldwell 78		6541	Globular Cluster	18 08	−43.7	CrA	6.6	13
Caldwell 79		3201	Globular Cluster	10 18	−46.4	Vel	6.8	18
Caldwell 80	ω Centauri	5139	Globular Cluster	13 27	−47.5	Cen	3.7	36
Caldwell 81		6352	Globular Cluster	17 26	−48.4	Ara	8.2	7
Caldwell 82		6193	Open Cluster	16 41	−48.8	Ara	5.2	15
Caldwell 83		4945	Galaxy	13 05	−49.5	Cen	8.6	20
Caldwell 84		5286	Globular Cluster	13 46	−51.4	Cen	7.6	9
Caldwell 85		IC 2391	Open Cluster	08 40	−53.1	Vel	2.5	50
Caldwell 86		6397	Globular Cluster	17 41	−53.7	Ara	5.7	26
Caldwell 87		1261	Globular Cluster	03 12	−55.2	Hor	8.4	7
Caldwell 88		5823	Open Cluster	15 06	−55.6	Cir	7.9	10
Caldwell 89	S Norma	6087	Open Cluster	16 19	−57.9	Nor	5.4	12
Caldwell 90		2867	Planetary Nebula	09 21	−58.3	Car	9.7	0.2
Caldwell 91		3532	Open Cluster	11 06	−58.7	Car	3.0	55
Caldwell 92	Eta Carina	3372	Emission Nebula	10 44	−59.9	Car	2.5	120
Caldwell 93		6752	Globular Cluster	19 11	−60.0	Pav	5.4	20
Caldwell 94	Jewel Box	4755	Open Cluster	12 54	−60.3	Cru	4.2	10
Caldwell 95		6025	Open Cluster	16 04	−60.5	TrA	5.1	12
Caldwell 96		2516	Open Cluster	07 58	−60.9	Car	3.8	30
Caldwell 97		3766	Open Cluster	11 36	−61.6	Cen	5.3	12
Caldwell 98		4609	Open Cluster	12 42	−63.0	Cru	6.9	5
Caldwell 99	Coalsack		Absorption Nebula	12 53	−63.0	Cru	—	350
Caldwell 100		IC 2944	Open Cluster	11 37	−63.0	Cen	4.5	15

Table A1.2. (continued)

Object	Name	NGC	Type	RA2000 H	m	Dec2000 degrees	Constellation	Visual* mag	Size** (')
Caldwell 101		6744	Galaxy	19	10	−63.9	Pav	8.4	16
Caldwell 102	Southern Pleiades	IC 2602	Open Cluster	10	43	−64.4	Car	1.9	50
Caldwell 103	Tarantula	2070	Emission Nebula	05	39	−69.1	Dor	8.2	40
Caldwell 104		362	Globular Cluster	01	03	−70.9	Tuc	6.6	13
Caldwell 105		4833	Globular Cluster	13	00	−70.9	Mus	7.4	14
Caldwell 106	47 Tucanae	104	Globular Cluster	00	24	−72.1	Tuc	4.0	31
Caldwell 107		6101	Globular Cluster	16	26	−72.2	Aps	9.3	11
Caldwell 108		4372	Globular Cluster	12	26	−72.7	Mus	7.8	19
Caldwell 109		3195	Planetary Nebula	10	10	−80.9	Cha	11.6	0.6

* This is the integrated magnitude over the whole area of the object. An angularly large object with a bright magnitude may therefore be less easy to see than a smaller object with a fainter magnitude. The magnitudes of the emission nebulae in particular may be misleading because they frequently contain brighter and darker regions. The symbol "≈" indicates a magnitude estimated from visual descriptions.

** This is the largest dimension of the object. Some objects may be filamentary or have a brighter core or outer region making them easier to see than might be expected.

(Data for this table obtained from *Sky Catalogue 2000.0*, Vol. 2 (Ed. A. Hirshfeld and R.W. Sinnott, Cambridge University Press, 1985); *The Caldwell Card* (Sky Publishing Corp., 1996); *NGC 2000.0* (R.W. Sinnott, Cambridge University Press, 1988); *Visual Astronomy of the Deep Sky* (R.N. Clark, Cambridge University Press, 1990); *Hartung's Astronomical Objects for Southern Telescopes* (D. Malin and D.J. Frew, Cambridge University Press, 1995); *Astrophysical Journal Supplement*, **4,** 257, 1959, S. Sharpless.)

The Greek Alphabet

Letter	Lower case	Upper case
Alpha	α	A
Beta	β	B
Gamma	γ	Γ
Delta	δ	Δ
Epsilon	ε	E
Zeta	ζ	Z
Eta	η	H
Theta	θ	Θ
Iota	ι	I
Kappa	κ	K
Lambda	λ	Λ
Mu	μ	M

Letter	Lower case	Upper case
Nu	ν	N
Xi	ξ	Ξ
Omicron	o	O
Pi	π	Π
Rho	ρ	P
Sigma	σ	Σ
Tau	τ	T
Upsilon	υ	Y
Phi	ϕ	Φ
Chi	χ	X
Psi	ψ	Ψ
Omega	ω	Ω

Bibliography

A3.1 Journals

Only the major and relatively widely available journals are listed. There are numerous more specialised research-level journals available in academic libraries.

Astronomy
Astronomy Now
Ciel et Espace
New Scientist
Scientific American
Sky and Telescope

A3.2 Ephemerises

Astronomical Almanac (published each year), H.M.S.O./U.S. Government Printing Office
Handbook of the British Astronomical Association (published each year), British Astronomical Association
Yearbook of Astronomy (published each year), Macmillan

A3.3 Star and Other Catalogues, Atlases and Reference Books

Astrophysical Quantities, CW Allen, Athlone Press, 1973
Burnham's Celestial Handbook, Vols 1, 2 and 3, R Burnham, Dover Press, 1978
Cambridge Deep-Sky Album, J Newton and P Teece, Cambridge University Press, 1983
Greek Myths, Vols 1 and 2, R. Graves, Penguin, 1966
Messier Album: An Observer's Handbook, JH Mallas and E Kreimer, Sky Publishing Corporation, 1978
Messier's Nebulae and Star Clusters, KG Jones, Cambridge University Press, 1991
Naked Eye Stars, R Lampkin, Gall & Inglis, 1972
Norton's 2000.0, I Ridpath (Ed.), Longman, 1989
Observing Handbook and Catalogue of Deep Sky Objects, C Luginbuhl and B Skiff, Cambridge University Press, 1990
Photographic Atlas of the Stars, H. Arnold, P Doherty and P Moore, IOP Publishing, 1997
Sky Atlas 2000.0, W Tirion, Sky Publishing Corporation, 1981
Sky Catalogue 2000, Vols 1 and 2, A Hirshfeld and R W Sinnott, Cambridge University Press, 1985
Star Tales, I Ridpath, Lutterworth Press, 1988
URANOMETRIA 2000.0, W Tirion, B Rappaport and G Lovi, Willman–Bell Inc., 1988

A3.4 Introductory Astronomy Books

Astronomy: A self-teaching guide, DL Moche, John Wiley & Sons, 1993
Astronomy: The Evolving Universe, M Zeilik, John Wiley & Sons, 1994
Astronomy: Principles and Practice, AE Roy and D Clark, Adam Hilger, 1988
Astronomy through Space and Time, S Engelbrektson, WCB, 1994
Introductory Astronomy and Astrophysics, M Zeilik, SA Gregory and EvP Smith, Saunders, 1992
Universe, WJ Kaufmann III, WH Freeman Publishers, 1994

A3.5 Practical Astronomy Books

Amateur Astronomer's Handbook, JB Sidgwick, Faber & Faber, 1971

Astronomical Telescope, BV Barlow, Wykeham Publications, 1975

Beginner's Guide to Astronomical Telescope Making, J Muirden, Pelham Press, 1975

Building and Using an Astronomical Observatory, P Doherty, Stevens Publications, 1986

Challenges of Astronomy, W Schlosser, T Schmidt-Kaler and EF Malone, Springer-Verlag, 1991

Compendium of Practical Astronomy, GD Roth, Springer-Verlag, 1993

Handbook for Telescope Making, NE Howard, Faber & Faber, 1962

Modern Amateur Astronomer, P Moore (Ed.), Springer-Verlag, 1995

Observational Astronomy, DS Birney, Cambridge University Press, 1991

Practical Astronomer, CA Ronan, Pan, 1981

Practical Astronomy with your Calculator, P Duffett-Smith, Cambridge University Press, 1981

Practical Astronomy: A User Friendly Handbook for Skywatchers, HR Mills, Albion Press, 1993

Seeing the Sky: 100 Projects, Activities and Explorations in Astronomy, F Schaaf, John Wiley & Sons, 1990

Seeing Stars, C Kitchin and RW Forrest, Springer-Verlag, 1997

Star Gazing through Binoculars: A Complete Guide to Binocular Astronomy, S Mensing, TAB, 1986

Star Hopping: Your Visa to the Universe, RA Garfinkle, Cambridge University Press, 1993

Telescopes & Techniques, C Kitchin, Springer-Verlag, 1995

Index

Index